Industrial Equipment for Chemical Engineering Set

coordinated by
Jean-Paul Duroudier

Heat Transfer in the Chemical, Food and Pharmaceutical Industries

Jean-Paul Duroudier

ISTE PRESS

ELSEVIER

First published 2016 in Great Britain and the United States by ISTE Press Ltd and Elsevier Ltd

ISTE Press Ltd
27-37 St George's Road
London SW19 4EU
UK

www.iste.co.uk

Elsevier Ltd
The Boulevard, Langford Lane
Kidlington, Oxford, OX5 1GB
UK

www.elsevier.com

British Library Cataloguing-in-Publication Data
A CIP record for this book is available from the British Library
Library of Congress Cataloging in Publication Data
A catalog record for this book is available from the Library of Congress
ISBN 978-1-78548-188-8

Contents

Preface

The observation is often made that, in creating a chemical installation, the time spent on the recipient where the reaction takes place (the reactor) accounts for no more than 5% of the total time spent on the project. This series of books deals with the remaining 95% (with the exception of oil-fired furnaces).

It is conceivable that humans will never understand all the truths of the world. What is certain, though, is that we can and indeed must understand what we and other humans have done and created, and, in particular, the tools we have designed.

Even two thousand years ago, the saying existed: "faber fit fabricando", which, loosely translated, means: "*c'est en forgeant que l'on devient forgeron*" (a popular French adage: *one becomes a smith by smithing*), or, still more freely translated into English, "practice makes perfect". The "artisan" (faber) of the 21st Century is really the engineer who devises or describes models of thought. It is precisely that which this series of books investigates, the author having long combined industrial practice and reflection about world research.

Scientific and technical research in the 20th century was characterized by a veritable explosion of results. Undeniably, some of the techniques discussed herein date back a very long way (for instance, the mixture of water and ethanol has been being distilled for over a millennium). Today, though, computers are needed to simulate the operation of the atmospheric distillation column of an oil refinery. The laws used may be simple statistical correlations but, sometimes, simple reasoning is enough to account for a phenomenon.

Since our very beginnings on this planet, humans have had to deal with the four primordial "elements" as they were known in the ancient world: earth, water, air and fire (and a fifth: aether). Today, we speak of gases, liquids, minerals and vegetables, and finally energy.

The unit operation expressing the behavior of matter are described in thirteen volumes.

It would be pointless, as popular wisdom has it, to try to "reinvent the wheel" – i.e. go through prior results. Indeed, we well know that all human reflection is based on memory, and it has been said for centuries that every generation is standing on the shoulders of the previous one.

Therefore, exploiting numerous references taken from all over the world, this series of books describes the operation, the advantages, the drawbacks and, especially, the choices needing to be made for the various pieces of equipment used in tens of elementary operations in industry. It presents simple calculations but also sophisticated logics which will help businesses avoid lengthy and costly testing and trial-and-error.

Herein, readers will find the methods needed for the understanding the machinery, even if, sometimes, we must not shy away from complicated calculations. Fortunately, engineers are trained in computer science, and highly-accurate machines are available on the market, which enables the operator or designer to, themselves, build the programs they need. Indeed, we have to be careful in using commercial programs with obscure internal logic which are not necessarily well suited to the problem at hand.

The copies of all the publications used in this book were provided by the *Institut National d'Information Scientifique et Technique* at Vandœuvre-lès-Nancy.

The books published in France can be consulted at the *Bibliothèque Nationale de France*; those from elsewhere are available at the British Library in London.

In the in-chapter bibliographies, the name of the author is specified so as to give each researcher his/her due. By consulting these works, readers may gain more in-depth knowledge about each subject if he/she so desires. In a reflection of today's multilingual world, the references to which this series points are in German, French and English.

The problems of optimization of costs have not been touched upon. However, when armed with a good knowledge of the devices' operating parameters, there is no problem with using the method of steepest descent so as to minimize the sum of the investment and operating expenditure.

1

Rotary Calcination Kiln: Heat Exchange by Radiation

1.1. General points

1.1.1. *Purpose of calcination kilns*

Calcination is the thermal dehydration of a metal hydroxide or of a salt. Cement kilns are not, technically speaking, calcination kilns, but the same technique is used.

1.1.2. *Specific characteristics of heat transfer*

In this case, heat is not transferred by convection, but rather by radiation; the wall of the kiln, which is heated using gas, transfers this heat to the product at points where direct contact occurs due to the rotation of the kiln. This regeneration effect may represent between 30% and 40% of the overall transfer of heat to the product. This transfer will be considered separately.

1.1.3. *Gas heating in kilns*

Dryers and kilns differ in terms of the temperature of the gas entering the assembly. If the gas is at a temperature lower than 400°C, the assembly is referred to as a dryer. If the gas is at a temperature above 400°C, the assembly is known as a calcination or roasting kiln.

The method used to heat the gas depends on the desired temperature. For temperatures up to 200°C, the gas is generally air, steam-heated using a finned tube exchanger. From 550°C to 1,500°C, smoke from burning natural gas, oil or powdered coal is used. The temperature may be adjusted to a point anywhere between 200°C and the fuel flame temperature by the introduction of cold air.

1.2. Description

1.2.1. *Basic principle of the kiln*

A rotary drum device (dryer or kiln) consists of a cylindrical shell turning on an axis (which is set at a slight inclination in relation to the horizontal plane). The divided material gradually moves down to the lower end of the kiln.

The shell is driven using a ring gear engaged with a motorized pinion. The drum is held in place using support treads and rollers.

Dryer shells consist of a simple steel sheet, generally fitted with lifting blades. These blades collect material, before scattering it through the gas current, improving contact between the solid and the gas, as we shall see below. The number of lifting blades in a dryer varies from 5 to 10 D (where D is the diameter of the shell, expressed in meters). However, the shells used in calcination kilns are lined with refractory bricks, which both provide thermal protection for the metal wall and insulate the hot gas/product mixture. This lining is usually around 0.15 m in thickness. The thermal conductivity of refractory bricks is generally of the order of 3.5 $W/m^{-1}/°C^{-1}$.

Generally, the diameter of the shell varies between 30 cm and approximately 4 m, i.e. between 2% and 20% of the length of the drum. The horizontal inclination varies from 0.02 to 0.06 m/m.

For safety reasons, the drum runs at a negative pressure of between 2 and 20 Pa, and a variety of equipment are used to maintain a complete seal between fixed and mobile elements.

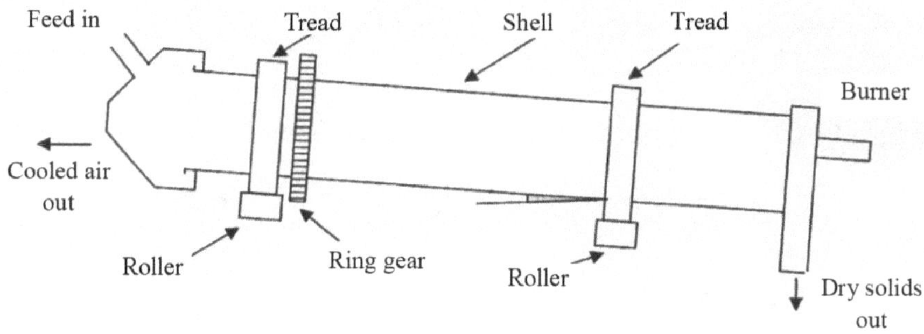

Figure 1.1. *Rotary kiln*

1.2.2. *Gas circulation in the kiln*

The bulk flow density of the gas varies between 0.3 and 5 kg of gas per second, per square meter of the cross-section of the kiln.

Gas may circulate in the same or opposite direction to the product.

At first glance, a counter-current system, in which gas circulates in the opposite direction to the product, might be expected to have a higher level of thermal efficiency than a co-current system, in which the gas and product circulate in the same direction. For this reason, the gas is always circulated in the opposite direction to the product in calcination kilns.

For drying purposes, however, a co-current gas flow may be used, which enables the use of hotter gas without any danger to the product. At high temperatures, less gas is required, reducing heat loss through the exhaust.

Furthermore, in certain cases, a co-current gas flow is strong enough to transport the smallest solid particles via a pneumatic effect. These particles are thus rapidly removed from the dryer, avoiding degradation.

In practical terms, to ensure that dust is collected before it has chance to settle, the speed of the gas in the exhaust conduit must be approximately 30 $m.s^{-1}$.

Inversely, we may wish to design a kiln in such a way that the finest particles will not be transported.

Considering that, in a turbulent regime, the weight of a particle balances out the air friction, we obtain the following maximum speed:

$$V_\ell = \sqrt{\frac{4}{3}\frac{d_p(\rho_s - \rho_a)g}{\rho_a C_x}}$$

For this problem, we shall take:

$$C_x = 0.8$$

Furthermore, considering ρ_a to be negligible in relation to ρ_s, we obtain:

$$V_\ell \# 4\sqrt{d_p \frac{\rho_s}{\rho_a}}$$

Or, in terms of the maximum value of the mass flow density of the gas:

$$\phi_{GM} = 4\sqrt{d_p \rho_s \rho_a}$$

ρ_a and ρ_s: bulk densities of the solid and air ($kg.m^{-3}$)

d_p: diameter of particles (m)

ϕ_{GM}: maximum flow density ($kg.m^{-2}.s^{-1}$)

EXAMPLE 1.1.–

$d_p = 400.10^{-6}$ m $\quad \rho_s = 1{,}200$ $kg.m^{-3}$ $\quad \rho_a = 1$ $kg.m^{-3}$

$$\phi_{GM} = 4\sqrt{400.10^{-6} \times 1200} = 2.77 \text{ kg.m}^{-2}.\text{s}^{-1}$$

The value used in practice is 60% of this limit, i.e.:

$$2.77 \times 0.6 = 1.6 \text{ kg.m}^{-2}.\text{s}^{-1}$$

1.2.3. *Support*

The support system, made up of treads and rollers, may be fine-tuned in a number of different ways. The kiln is set at an inclination in relation to the horizontal plane, and to avoid movement, the rollers are adjusted along an axis set at an angle α in relation to the drum axis.

In the diagram below, the drum is prevented from slipping to the right by the angle of the rollers.

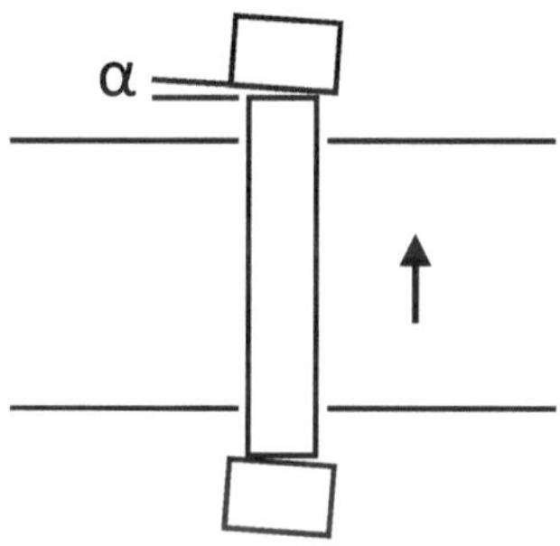

Figure 1.2. *Roller adjustment*

In large kilns, further support is provided by hydraulic buffers that move additional rollers. The thrust from these buffers, running parallel to the axis of the kiln on alternate sides, creates a vertical raising and lowering treads movement, which evens out wear on the treads and support rollers.

Either a *floating* or, more often, an adjusted tread may be used; the two types differ in the way in which they are fixed to the kiln shell.

Floating tires can be repositioned, i.e. placed opposite to the support rollers in order to compensate for the fact that the length of the cylinder increases due to dilation when the kiln is heated.

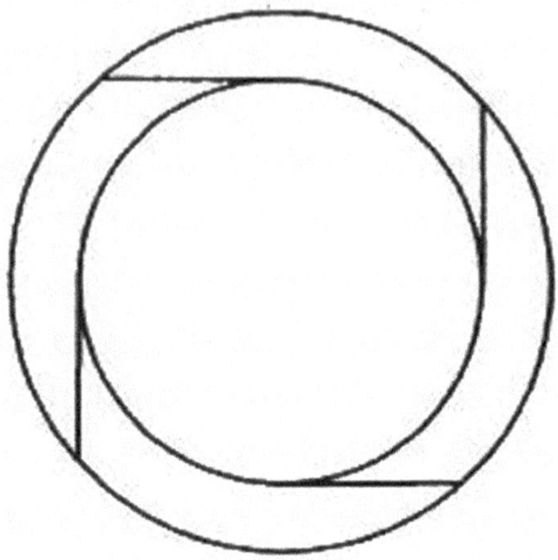

Figure 1.3. *Floating type metal tread*

Adjusted treads cannot be repositioned in this way, so the support rollers must be able to move. Adjusted treads reduce ovalization in the cylinder, a phenomenon that damages the internal coating.

These elements are costly, and it is best to limit the number used in a support system as far as possible.

1.2.4. *Lining*

Calcination kilns are generally lined. This lining acts both as thermal insulation, keeping heat in, and to protect the cylinder from the high-temperature gases which are used.

This lining is made from silica or magnesia bricks. Magnesia, the cheapest option, has a slightly higher dilation level when heated, and tends to creep under external loads at temperatures above 1,500°C.

In cases where the temperature of the gas entering the system does not exceed 750°, lining is not essential as long as the cylinder is made from heat-resistant steel. If ordinary steel is used, the maximum acceptable temperature is 400°C (for dryers); in this case, the cylinder is insulated from the outside (using fiberglass).

1.2.5. *Seals at drum ends*

Joints which produce a fully gas-tight seal have a low life expectancy and drain engine power. Joints with a higher leakage level have a longer life expectancy. No manufacturer has yet produced a seal combining low leakage and long life expectancy. Currently, a life expectancy of 1 year is considered satisfactory (i.e. between two programmed down times). A variety of seal types are available, for example, using air cooling.

In practice, dryers are operated at a low level of negative pressure (2–20 Pa) in order to avoid smoke and dust leakage into the workspace.

1.2.6. *Kinematic drive train in calcination kilns*

The kiln can be moved, at normal speed, by the main motor, or at slower speeds by a turning motor, which serves two purposes:

1) When the kiln is cold, the turning motor can be used to move the kiln into the best position for maintenance work.

2) The turning motor starts automatically when the main motor stops, in order to prevent the shell from warping due to its own weight and the weight of the solid material inside.

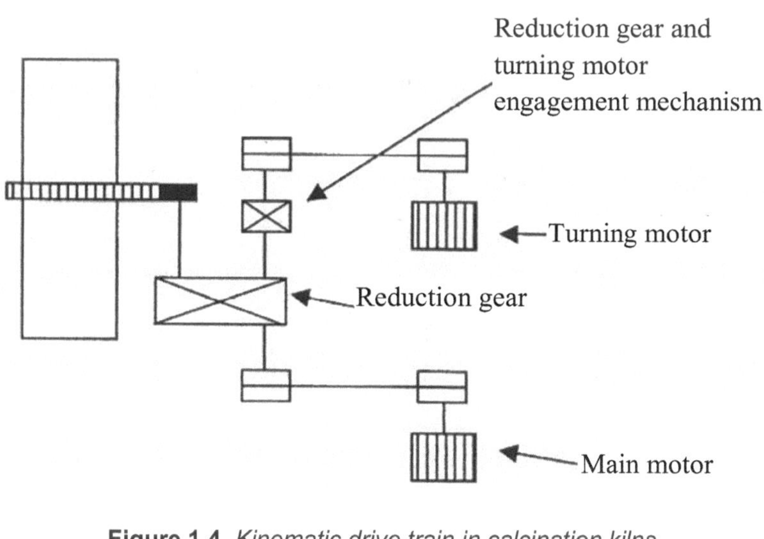

Figure 1.4. *Kinematic drive train in calcination kilns*

When the main motor is operational, the turning motor is automatically disengaged. Turning motor systems are particularly important when working at high temperatures (calcination).

1.2.7. *Feeding a rotary dryer*

Rotary dryers may be fed using a variety of types: sprayed liquids, slurries, and wet or sticky fuel pastes.

In order to improve feed flow, screws may be placed inside the first meter or meters of the shell, with a pitch of around 1 m. If the feed is particularly sticky, the shell may be equipped with hammers, placed on the outside of the cylinder. Another option is to attach chains to the lifter blades. However, hammers are noisy, and chains increase wear on the kiln lining.

In order to make the feed supply more uniform, it is sometimes necessary to add some of the dry products produced by the kiln to the raw material. A recycling system designed to fulfill this function includes:

- a dry product conveyer;
- a storage hopper for the dry product;
- a feed system (screw, vibratory feeder, etc.);
- a paddle-type mixer, sometimes followed by a disc or a hammer mill.

Recycling in this way is only possible if the product in question is able to withstand prolonged exposure to high temperatures.

1.2.8. *Grain breaking*

Within the drum, fine particles are created by breaking up grains. This may be done by a shock effect (with grains falling from a height), by wearing the grains down (through rubbing against other grains and the walls of the kiln) or by explosion, when the water inside the grain evaporates; this is particularly common in cases where a chemical bond exists.

1.2.9. *Shaft power of the electric motor*

The power required is the sum of the following two terms:

– a term P_P directly linked to the presence of the product;

– a term P_F which represents friction between mechanical elements.

1) To maintain the product in an inclined position (calcination kiln):

The mass of the product present in the kiln is:

$$R\frac{\pi D^2}{4}L\rho_a$$

The lever will be assimilated to its maximum value $\frac{D}{2}\sin\beta$.

The angular velocity is $2\pi N$, where N is the frequency of rotation. The coefficient R is the fill ratio in the shell.

This gives us a corresponding power (in Watts):

$$P_P = \frac{\pi^2 D^3}{4} N \sin\beta R L \rho_a g$$

β: angle formed between the surface of the product and the horizontal plane. This angle is very close to the angle of repose of the solid, which is often approximately 33°.

2) To mix and lift the product using lifting blades:

With each turn of the shell, the mass M_P of the product present is lifted to a height equal to the diameter of the drum. Theoretically, therefore:

$$P_P = N\,D\,M_P\,g$$

This value is increased by 50% to take account of product mixing, and of the fact that only a small proportion of the product will be found in an inclined position in the lower part of the drum. Hence:

$$P_P = 1.5\,N\,D\,M_P\,g$$

3) Including mechanical friction:

$$P_F = 4.5\times10^{-4}\left(M_P + M_E\right)gV_p$$

M_P and M_E: mass of product and mass of empty equipment (kg)

g: acceleration due to gravity (9.81 $m.s^{-2}$)

V_p: peripheral velocity of rollers ($m.s^{-1}$)

$$V_p = N\pi\left(D + 0.5\right)$$

The shaft power of the electric motor is thus:

$$P_a = \frac{P_P + P_F}{\eta}$$

η: output of the moto-reducer system

EXAMPLE 1.2.–

In the case of a dryer:

$D = 2.7$ m $\quad M_E = 20{,}000$ kg

$N = 0.1$ rev.s^{-1} $\quad R = 0.1$

$L = 7$ m $\quad \eta = 0.92$

$\rho_a = 1{,}600$ kg.m^{-3}

$$M_P = 0.1\times\frac{\pi\times2.7^2}{4}\times7\times1600 = 6{,}412 \text{ kg}$$

$$V_p = \pi\left(2.7 + 0.5\right)\times0.1 = 1 \text{ m.s}^{-1}$$

$$P_a = \frac{1}{0.92}\left(1.5\times0.1\times2.7\times6{,}412\times9.81 + 4.5\times10^{-3}\left(20{,}000 + 6{,}412\right)\times9.81\times1\right)$$

$$P_a = 28{,}958 \text{ W} = 29 \text{ kW}$$

1.2.10. *Axial transport of products in a bare drum in the absence of gas*

This phenomenon was discussed in detail by Seaman [SEA 51] and then by Hogg *et al.* [HOG 74]. We shall take a different approach.

The circumferential (orthoradial) speed at each point of the shell is NπD, where N is the drum rotation frequency and D its diameter.

From an external perspective, the solid does not appear to turn, but rather slides inside the drum, so that the orthoradial speed of the solid in relation to the drum is equal and opposite to the previous speed.

At the center of the solid mass M, we have:

$\overrightarrow{MV}$ = the average speed of the solid in relation to the drum;

$\overrightarrow{MU}$ = the orthoradial speed of the solid in relation to the drum;

$\overrightarrow{MW}$ = the axial speed of the solid in relation to the drum:

$$\overrightarrow{MV} = \overrightarrow{MU} + \overrightarrow{MW}$$

By definition, $\overrightarrow{MU}$ and $\overrightarrow{MW}$ are perpendicular.

Now, let us introduce an angle α between the average speed $\overrightarrow{MV}$ of the solid in relation to the drum and any plane perpendicular to the drum axis. The model of $\overrightarrow{MW}$ is thus:

$$MV = MU \text{tg}\alpha = N\pi D \text{tg}\alpha$$

During the time τ in which the product is present in the drum, this product covers an axial distance L equal to the length of the drum:

$$L = \tau N\pi D \text{tg}\alpha$$

$$\tau = \frac{L}{N\pi D \text{tg}\alpha}$$

We must now determine the angle of progression α which, contrary to what one might think, is different to the angle υ formed by the slope of the drum and the horizontal plane.

1.2.11. *Determining the angle of progression α*

Around point M, the center of the solid mass, the shell wall can be assimilated to an inclined plane along which the product runs in such a way that the free surface creates an angle β with the horizontal plane; by definition, this angle is equal to the natural flow angle of the product.

In the vicinity of point M, the free surface of the solid is parallel to a plane tangential to the drum, such that the line between the point of contact of the plane and the point M crosses the drum axis.

Taking P as the weight of the solid present in the drum over one unit of length (1 m), the component of the weight parallel to the free flow surface is P sin β. The average speed MV of the solid is thus (modulus):

$$MV = K_v P \sin\beta$$

Coefficient K_V is used to account for the fact that the speed profile is a sigmoid.

The axial component MW of the mean speed of the solid is proportional to component Psinν of the weight P in relation to the drum axis, hence (again, using a modulus value):

$$MW = K_a \times P \sin\nu$$

According to the definition of angle α:

$$\sin\alpha = \frac{MW}{MV} = \frac{K_a}{K_v}\frac{\sin\nu}{\sin\beta}$$

The relationship K_a/K_v is less than 1. The average value of K_a/K_v is considered to be equal to 0.65.

1.2.12. *Expression of time spent in the drum*

Angle α has a value of a few hundredths of a radian, so we can write:

$$tg\alpha \# \sin\alpha$$

This gives us the following expression of time spent in the drum:

$$\tau = \frac{1.53 L \sin\beta}{N\pi D \sin\nu} \qquad (1.53 = 1/0.65)$$

EXAMPLE 1.3.–

$\beta = 35°$ (sin 35° = 0.57) $N = 0.16$ rev.s^{-1}

$\nu = 0.02$ rad $D = 0.7$ m

$L = 5$ m

$$\tau = \frac{1.53 \times 5 \times 0.57}{0.16 \times \pi \times 0.7 \times 0.02} = 619.60 \text{ s}$$

$$\tau = 10 \text{ mn } 20 \text{ s}$$

1.2.13. *Accounting for the presence of gas*

In the absence of lifting blades, the gas current only transports a tiny proportion (by weight) of fine particles, and the expression used above remains valid.

If lifting blades are used, however, particles fall into the gas current, and the time spent in the drum needs to be corrected using a factor Δτ.

Friedman and Marshall give the following approximate expression of Δτ:

$$\Delta\tau = \pm \frac{9.84 W_G L}{d_P^{0.5} W_P}$$

The + sign corresponds to counter-current circulation and the – sign to co-current circulation.

EXAMPLE 1.4.–

$D_p = 200\ \mu m \quad W_P = 0.05\ kg/s^{-1}$

$L = 5\ m \qquad W_G = 0.25\ kg/s^{-1}$

$$\Delta\tau = \pm\frac{9.84 \times 0.25 \times 5}{(200)^{0.5} \times 0.05} = \pm 17s$$

NOTE (Peripheral Velocity).–

Authors often link the peripheral speed of a calcination kiln to the nature of the product being processed.

Calcination is a chemical decomposition reaction, and for the process to be complete, the product needs to remain in the reactor drum for a certain length of time.

As calcination systems do not include lifting blades, the stay time is expressed as (see section 1.2.12):

$$\tau = \frac{1.53\ L\sin\beta}{N\pi D\sin v}$$

The angle v of the slope of the drum varies between 0.02 and 0.06 rad, giving a mean value of 0.04 radians. Furthermore, system N π D is equal to the peripheral velocity V_P of the shell (V_P in $m.s^{-1}$ and N in $rad.s^{-1}$).

This gives us the following expression of peripheral velocity for a given calcination reaction and a given kiln length:

$$V_P = \frac{1.53\ L\sin\beta}{\tau \times 0.04} \qquad (m.s^{-1})$$

For example:

Processed product	V_P ($m.mn^{-1}$)
Titanium oxide	1
Cement	15
Phosphate	40

1.2.14. *Fill ratio*

Given the solid product flowrate Q_P (in m^3 appearing per second) through the system, i.e. the system's production capacity, the average stay time τ and the volume V of the drum, we are able to find the fill level R, or the solid holding capacity:

$$R = Q_P \tau / V$$

Moreover, in the absence of lifting blades, the solid can be seen to occupy a surface of the cross-section of the drum bounded by a circle arc and the chord which forms it. Taking 2α as the angle at the center which subtends the circle arc, the fill level is written:

$$R = \alpha - \frac{1}{2}\sin 2\alpha$$

The fill level is generally between 5% and 15%.

EXAMPLE 1.5.–

$$\alpha = \frac{\pi}{6}\,\mathrm{rad} = 30°$$

$$R = 0.523 - 0.433 = 0.09$$

It can be found that 9% of the drum volume is occupied by the apparent volume of the solid.

1.3. Heat transfer by radiation: basic concepts

1.3.1. *Working hypothesis*

In this case, we will only consider a configuration in which gas is contained in a fully *concave* container, the walls of which may be divided into several parts, each at a different temperature. Furthermore:

– the gas is at a uniform temperature T_G, and is presumed *gray*, i.e. its physical properties in terms of radiation (emissivity and absorptivity) are independent of the wavelength spectrum of the radiation, and are solely dependent on the composition of the gas (water vapor and carbon dioxide content);

– each element of the wall is also gray and has a uniform temperature T_I, which is distinct from the temperature of the other elements. One of these elements is the free (or gas contact) surface of the product, which may be:

- molten metal;

- ceramic (either soft-paste porcelain fired at 900°C, or hard-paste porcelain, fired at 1,300°C);

- a divided solid: this is the case in cement kilns, and in calcination kilns as a whole. Etymologically speaking, calcination comes from the Latin *calcinare*, meaning to burn limestone (calcium carbonate) in order to produce calcium oxide (CaO, quicklime) by a rise in temperature.

By extension, this term is applied to the dehydration of any metal hydroxide.

We shall not give detailed consideration to combustion chambers, such as those found in most systems of the following types:

– steam boilers (oil, coal or gas-fired);

– incineration furnaces (which, etymologically speaking, reduce a product to cinders) for garbage or waste products;

– petrol/oil furnaces (for refining and cracking).

A detailed study of combustion chambers would require consideration of the following elements:

– the structure and emissivity of the flame (with the potential presence of solid particles);

– breakdown of the volume of gas into distinct zones of more or less uniform temperature.

These considerations lie outside the sphere of process engineering and belong to the field of mechanical and thermal engineering.

1.3.2. *Key concepts*

The *black* emittance of a surface is:

$$E = \sigma T^4$$

σ: Stéfan's constant (5.67×10^{-8} W.m^{-2}.K^{-4})

T: absolute temperature (K)

The emittance E is then measured in W.m^{-2}.

The emissivity ε of a real surface (solid or liquid) is the relationship between the real emittance and the black emittance, and the emittance of any given surface may be expressed as:

$$\varepsilon E = \varepsilon\sigma T^4$$

The reflexivity ρ_S (or the reflection coefficient) of a real surface is the ratio between the reflected radiation energy and the incident radiation energy.

Expressed in terms of flow density:

$$\rho_s = \frac{\phi_r}{\phi_1}$$

ϕ_1 : incident energy flow.

The absorptivity of a solid surface is thus:

$$\alpha_s = 1 - \rho_s = \frac{\phi_a}{\phi_1}$$

The emissivity of a gas toward a solid or liquid surface with which it is in contact is defined in the same way, and the emittance of the gas:

$$\varepsilon_g E_g = \varepsilon_g \sigma T_G{}^4$$

The transmissivity ε_τ of a gas crossed through by radiation over a thickness of r is:

$$\exp\left[-\int_0^r K(r)\,dr\right] \# \exp\left[-\bar{K} \times r\right]$$

$\bar{K}$ is the attenuation coefficient measured in m^{-1}.

The transmissivity is given by the following relationship:

$$\tau_g \frac{\text{energy transported after crossing through gas}}{\text{energy transported before crossing through gas}}$$

The average coefficient $\bar{K}$ (over distance) has been shown, experimentally, to depend on the temperature of the gas and on the pressure P_G of the gas (if the gas is a pure body):

$$\tau_g\left(T_G P_{G'} r\right) = \exp\left[-\bar{k}\left(T_{G'} r\right) P_G r\right]$$

The absorptivity of a gas under thickness r is:

$$\alpha_g = 1 - \tau_g$$

In practice, a gas may be considered to absorb or transmit without reflecting.

However, the surfaces under consideration here are opaque, and may thus reflect or absorb without transmitting:

$$\alpha_s = 1 - \rho_s$$

1.3.3. *Kirchoff's law for surfaces*

In an isothermal container in a state of thermal equilibrium, let us consider n bodies with an external surface A_i (with i from 1 to n). As the container is in a state of equilibrium, the energy flow density through any imaginary surface within the container is equal to E_0 following Plank's expression, i.e. everywhere has the same value.

Taking α_i to represent the absorptivity of the n bodies, each receives the following flow:

$$\phi_{ai} = E_0 A_i \alpha_i$$

As the system is in equilibrium, the heat flow emitted by each body is equal to the received heat flow:

$$\phi_{ei} = E_0 A_i \varepsilon_i \quad \text{and} \quad \phi_{ei} = \phi_{ai}$$

Thus, in the absence of heat transfer:

$$\varepsilon_i = \alpha_i$$

The emissivity ε_i is a state variable for solids and liquids, and is solely dependent on the nature of the body and its temperature.

Kirchhoff's identity remains valid in the presence of heat transfer as, if the surface of a condensed body (i.e. a solid or a liquid) is in contact with a medium of a different temperature, the resulting temperature disturbance will not affect much of the body below surface level, and evaluation of ε_i at temperature T_S of the solid or liquid body still provides sufficiently precise results.

As we shall see, Kirchhoff's law is not valid for gases, which have a thermal diffusivity ($\lambda/C\rho$) approximately 1,000 times higher than that of a solid or a liquid. In the presence of heat transfer, when evaluating α_g, consideration must be given to both the overall temperature T_G of the gas and the temperature T_S of the surface in contact with the gas.

1.3.4. *Luminance and emitted flow density*

Consider a point source of electromagnetic radiation located at point O_l, with $d^2\phi_l$ the energy transported per second along a beam with a straight section of $d\sigma_2$ centered on point O_2 at a distance of r_{12} from point O_l. We write:

$$d^2\phi_{1\to 2} = dl_1 d\omega_1$$

O_l and $d\sigma_2$ are considered to exist in a vacuum and in isolation, i.e. we will only consider the direct exchange between O_l and σ_2.

$d\omega_l$ is an elementary solid angle:

$$d\omega_1 = \frac{\sigma_2}{(r_{12})^2}$$

dI_l is the energy intensity of the beam of electro-magnetic radiation.

However, point sources are merely an abstract notion, and in reality, radiation cannot only be emitted from an arbitrarily small elementary surface dA_l. We can therefore write, *a priori*:

$$dI_1 \propto J dA_1$$

$d\sigma_l$ is the apparent area of the surface dA_l seen from the point of reception. In addition:

$$d\sigma_1 = dA_1 \cos\theta_1 \quad (\theta_1\text{: angle between } O_1O_2 \text{ and between the normal and } dA_1)$$

Similarly, if the emitting surface is dA_2, we obtain:

$$d\sigma_2 = dA_2 \cos\theta_2$$

We may also suppose that dA_2 emits toward dA_l, hence the two expressions:

$$d^2\phi_{1\to 2} = dI_1 d\omega_1 \quad \text{and} \quad d^2\phi_{2\to 1} = dI_2 d\omega_2$$

Each of these flows is the sum of two terms:

– a source term, dependent solely on the temperature of the emitting surface;

– a term accounting for the reflection of incident radiation;

In other words, the energy intensity values are dependent on the temperatures T_l and T_2 of surfaces dA_l and dA_2.

$$dI_1 = dI_1(T_1, T_2) \text{ and } dI_2 = dI_2(T_2, T_1)$$

Let us now suppose that a state of thermal equilibrium has been established, i.e.:

$$T_1 = T_2 = T$$

In this case, there is no energy transfer between dA_1 and dA_2. In other words, the previous flows take the following form:

$$d^2\phi_{e1\to 2} = d^2\phi_{e2\to 1}$$

$$dI_1 d\sigma_2 = dI_2 d\sigma_1$$

$$\frac{dI_1}{d\sigma_1} = \frac{dI_2}{d\sigma_2} = L(T_1, T)$$

Whatever the known temperature T, dI_1 and dI_2 are thus proportional to $d\sigma_1$ and $d\sigma_2$ respectively. We will assume that, if temperatures T_1 and T_2 change, this proportionality is maintained, and we may write:

$$d^2\phi_{e1\to 2} = L_1(T_1, T_2)\frac{d\sigma_1 d\sigma_2}{r^2} \qquad [1.1]$$

$$d^2\phi_{e2\to 1} = L_1(T_2, T_1)\frac{d\sigma_1 d\sigma_2}{r^2} \qquad [1.2]$$

Now, let us introduce the angles θ_1 and θ_2 between the normals to dA_1 and dA_2 and the straight line O_1 O_2. Moreover:

$$\sigma_1 = dA_1 \cos è_1 \qquad \sigma_2 = dA_1 \cos è_2$$

$$d\omega_1 = \frac{\sigma_2}{r^2} \qquad d\omega_2 = \frac{\sigma_1}{r^2}$$

Thus:

$$d^2\phi_{e1\to 2} = L_1 \cos\theta_1 dA_1 d\omega_1$$

$$d^2\phi_{e2\to 1} = L_2 \cos\theta_2 dA_2 d\omega_2$$

Regarding the flow densities:

$$\varphi_{1\to 2} = \frac{d\phi_{1\to 2}}{dA_1} \text{ and } \varphi_{2\to 1} = \frac{d\phi_{2\to 1}}{dA_2}$$

$$d\varphi_{1\to 2} = L_1 \cos\theta_1 d\omega_1$$

$$d\varphi_{2\to 1} = L_2 \cos\theta_2 d\omega_2$$

(These two equations reflect Lambert's cosine law).

To obtain the omnidirectional emitted flow densities φ_{e1} and φ_{e2}, we simply carry out an integration using the solid angle:

$$\varphi_{el} = \int_{\omega_1=0}^{\omega_1=2\pi} d(\phi_{1\to 2}) = L_1 \int_0^{2\pi} \cos\theta_1 dw_1$$

$$\varphi_{el} = L_1 \int_{\beta_1=0}^{2\pi} d\beta_1 \int_{\theta=0}^{\theta=\frac{\pi}{2}} \cos\theta_1 \sin\theta_1 d\theta_1 = \pi L_1$$

θ_1: angle with the y-axis

β_1: angle with the x-axis.

Lambert's cosine law means that the energy intensity of an elementary light beam emitted by an elementary surface dA_1 is proportional to the apparent area $d\sigma_1$ of the emitting surface seen from the receiving surface. The proportionality coefficient is the *luminance*, L_1, of the emitting surface, and the relationship between the luminance and the emitted omnidirectional flow density φ_{cl} is as follows:

$$\varphi_{el} = \pi L_1$$

1.3.5. *Specular reflection and diffuse reflection*

1) Purely specular reflection:

Any ray reflected from point P_i of a specular surface (mirror) toward a point P_j of a surface dA_i appears to come from a virtual source situated in an extension of the straight line P_jP_i and symmetrical to the real source in a plane tangential to surface A.

The intensity of a light beam of this type is:

$$dI_i = L_i d\sigma_i$$

$d\sigma_i$ is the apparent area of the virtual source as observed from point P_j. When the ray P_iP_j describes 2 π steradians at point P, it appears to come from the whole of the space located behind the plane tangential to dA_i, and its luminance is dependent on the sources present in the real space; it is totally *independent* of the nature and the temperature of the reflecting surface.

2) Purely diffuse reflection:

The luminance of any ray reflected from a point P_i of a reflective surface is independent of the direction P_iP_j of the receiving point. Hence, if φ_{li} is the density of the flow incident on the element of surface dA_i, then the reflected flow density is:

$$\varphi_{ri} = \rho_i \varphi_{li}$$

Even if the incident radiation does not have the same luminance in all directions, *the reflected radiation is isotropic*, and we have:

$$dl_{ri} = L_{ri} d\sigma_i = \frac{\rho_i \varphi_{li}}{\pi} d\sigma_i$$

The luminance L_{ri} is not dependent on the angle formed by the direction in question with the normal to dA_i.

1.3.6. *Radiosity of a perfect exchange surface*

Radiosity is the sum of gray emittance and gray reflectance:

$$J_i = \varepsilon_i \sigma T_i^4 + \rho_i \varphi_{li}$$

However, due to the presence of the incident flow density φ_{li}, radiosity is not an intrinsic property of the surface in question; it depends on the geometric configuration of the container and on the thermal state of all of its elements. For this reason, it is better to simply use the term "emitted flow density", and to write:

$$\varphi_{ei} = \varepsilon_i \sigma T_i^4 + \rho_i \varphi_{li}$$

1.3.7. *Equivalent direct transfer surface*

For two isolated elementary surfaces in a vacuum:

$$d^2\varphi_{i \to j} = L_i \frac{d\sigma_1 d\sigma_2}{r_{ij}^2} = \varphi_{ei} \frac{d\sigma_i d\sigma_j}{\pi r_{ij}^2}$$

The equivalent elementary direct transfer surface is:

$$d^2\overline{s_i s_j} = \frac{d\sigma_i d\sigma_j}{\pi r_{ij}^2}$$

Moreover, for two finite surfaces A_1 and A_2 in a vacuum, the equivalent direct transfer surface is:

$$\left(\overline{s_i s_j}\right)_0 = \int_{A_1}\int_{A_2} \frac{\cos\theta_i dA_i \cos\theta_j dA_j}{\pi r_{ij}^2}$$

In the presence of gas along the trajectory r_{ij}, only the fraction τ (r_{ij}) is transmitted from the element of surface dA_i to the element dA_j. Fraction τ (r_{ij}) is the transmissivity of the gas along the trajectory r_{ij}, and this quantity is symmetrical in relation dA_i and dA_j.

Thus, in the presence of gas:

$$\overline{s_i s_j} = \int_{A_1}\int_{A_j} \frac{dA_i \cos\theta_i dA_j \cos\theta_j \tau\left(r_{ij}\right)}{\pi r_{ij}^2} \neq \overline{s_j s_i}$$

The precise calculation of equivalent direct transfer surfaces is extremely complicated, as these surfaces no longer depend solely on the geometric configuration of the container, but also on the nature of the gas. Moreover, the direction of propagation of the radiation should also be taken into account, although this is not included in the equations given above.

1.3.8. *Equivalent trajectory and average trajectory through a gas*

Equivalent surfaces in the presence of gas become much easier to calculate if we remove the transmissivity from the integrand, writing:

$$\overline{s_i s_j} = \tau(L_m) \times \left(\overline{s_i s_j}\right)_{vacuum}$$

L_m is the "average trajectory" of the radiation. This trajectory L_m depends on both the size and the shape of the container in question. The relationship given above is only valid for a specific size of the container, and if different values are used, we must write:

$$\overline{s_i s_j} = \phi\tau(L_m) \times \left(\overline{s_i s_j}\right)_{vacuum}$$

However, if a value for L_m is selected within the interval of usual industrial values, Hottel *et al.* [HOT 67] note that the *correction coefficient* ϕ will be no more than 10% different from the value of 1. The average trajectory L_m is deduced from the equivalent trajectory L_e.

The equivalent trajectory at a point P of the wall of a container is the radius of an imaginary gaseous hemisphere centered on P, which produces the same energy flow as the total gaseous volume of the container at this point.

The size of a container of a given shape (cylinder, parallelepiped, sphere, etc.) is characterized by a characteristic dimension D. Moreover, the transmissivity is an exponential function of product Kr. In this product, the distance r is replaced by the characteristic dimension D. Furthermore, K is the attenuation coefficient.

The relationship L_e/D is a rapidly decreasing function of the product KD. Thus, for a hollow sphere, when KD changes from zero to 10, L_e/D goes from 0.67 to 0.4. In practice, however, KD is considerably lower than 10, and we can write:

$$\alpha_g = 1 - \tau_g = 1 - e^{-KL_e} \# KL_e$$

In a state of thermal equilibrium:

$$\varepsilon_g = \alpha_g = KL_e$$

If A is the wall surface of the container holding the gas, then the flow emitted by the gas toward the wall is:

$$\phi_{gP} = \varepsilon_g E_g A = KL_e E_g A$$

However, the omnidirectional energy flow absorbed by a volume V of gas has been shown [HOT 67] to be:

$$\phi_{ga} = 4KE_g V$$

Considering these two flows to be equal, we obtain:

$$L_e = \frac{4V}{A}$$

Hottel *et al.* [HOT 67] used the results of numerous calculations to show that the L_m/L_e relationship is, on average, approximately 0.88, so that:

$$L_m \# \frac{3.5V}{A}$$

In this expression, the gas is considered to be at a uniform temperature T_G, and the inside of the container is considered to be fully concave.

Note that, for a cylinder of diameter D and infinite length:

$$L_m \neq 0.94D. \quad [1.3]$$

Hottel *et al.* [HOT 67, p. 266] provide variation curves for $\overline{ss}$ and $\overline{gs}$ (for plates, cylinders and spheres) as a function of KD.

1.3.9. *Emissivity, absorptivity and transmissivity of the gaseous mixture (calculation)*

Our discussion in this section will be limited to combustion vapors. These mixtures only contain two significant components: carbon dioxide and water vapor. Furthermore, we shall consider that, as in the case of most combustion reactions, the total pressure is similar to that of the atmosphere.

1) Emissivity

Consider the emissivity of a gaseous mixture radiating *toward* the wall. The known values include:

- the temperature TG of the gaseous mixture;
- the characteristic average trajectory L_m of the container;
- the partial pressures P_C and P_V of carbon dioxide and water vapor.

Figures 1.5 and 1.6 provide values for ε_c (CO_2) and ε_v (H_2O) as a function of the temperature T_G of the gaseous mixture.

These curves are parametered by the products $p_c L_m$ and $p_v L_m$. From this, we obtain the emissivity of the mixture:

$$\varepsilon_g = \varepsilon_c + \varepsilon_v - \Delta\varepsilon$$

Using a value of approximately 0.03 for $\Delta\varepsilon$ provides satisfactory results, but more precise curves have been published by De Vriendt [DE 84, p. 231]. Note that ε_g is not dependent on the receptor surface (see Figures 1.5 and 1.6).

2) Absorptivity and transmissivity

The absorptivity considered here is that of the gaseous mixture, absorbing a flow *emanating* from a surface with an absolute temperature T_S different from that of the gas, T_G. First, let us calculate the following systems:

$$P_c L_m \frac{T_S}{T_G} \quad \text{and} \quad P_v L_m \frac{T_S}{T_G} \qquad [1.4]$$

The curves shown in Figures 1.5 and 1.6 provide values of ε_c and ε_v for the temperature T_S. The absorptivity of the gas is:

$$\alpha_{gS} = \varepsilon_c \left(\frac{T_G}{T_S}\right)^{0.65} + \varepsilon_v \left(\frac{T_G}{T_S}\right)^{0.45} - \Delta\varepsilon \qquad [1.5]$$

$\Delta\varepsilon$ may be estimated at a value of 0.03.

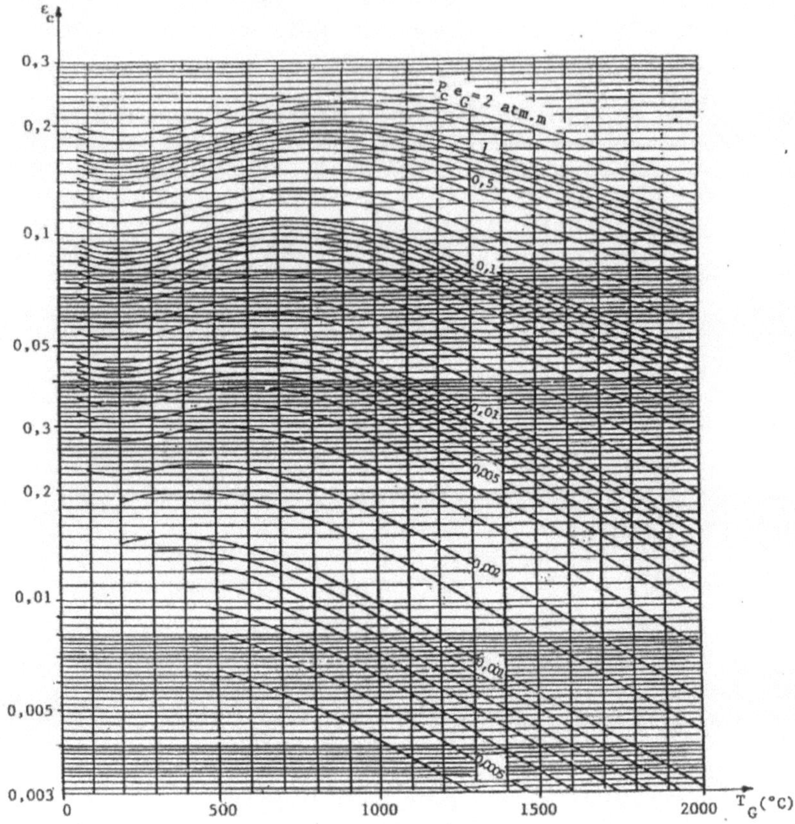

Figure 1.5. *Emissivity of carbon dioxide [DE 84]*

Finally, the transmissivity of the gas for radiation *emanating* from the wall element A_i at a temperature T_i and after a trajectory L_m to reach surface A_j is written:

$$\tau_{ij} = 1 - \alpha_i \qquad [1.6]$$

This transmissivity is dependent on the temperature of the gas and that of the emitting surface A_i. The receptor surface A_j is only relevant insofar as it affects the length L_m.

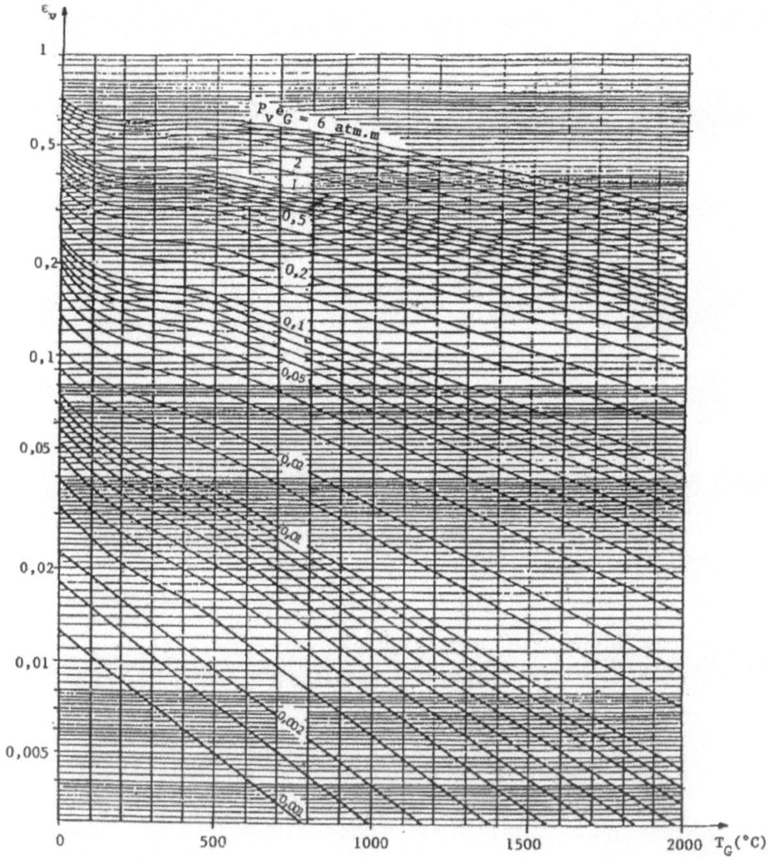

Figure 1.6. *Emissivity of water [DE 84]*

1.3.10. *Distribution of direct radiation*

According to our hypotheses, the container under consideration is fully concave on the inside, meaning that each point on this surface can "see" all elements of the internal wall. Let A_T be the total surface of this wall.

More precisely, in the absence of gas, we can write that, for all points P_i:

$$\sum_{i=l}^{n}\left(s_i s_j\right)_{vacuum} = \int_{wall\ A_T} d^2\left(s_i s_j\right)_{vacuum}$$

We know that:

$$d_{ij}^2\left(\overline{s_i s_j}\right)_{vacuum} = \frac{d\sigma_i d\sigma_j}{\pi\left(r_{ij}\right)^2} = dA_i \frac{\cos\theta_i d\omega_1}{\pi}$$

And also:

$$\int_{\omega_i=0}^{\omega_i=2\pi} \cos\theta_i d\omega_i = \int_0^{\frac{\pi}{2}} d\beta_i \int_{\theta_i=0}^{\pi} \cos\theta_i \sin\theta_i d\theta_i = \pi$$

Sweeping the solid angle 2π from dA_i, we encounter all of the surfaces dA_j visible from dA_i:

$$\sum_j d_{ij}\left(\overline{s_i s_j}\right)_{vacuum} = \int_{\omega_i=0}^{\omega_i=2\pi} d_{ij}^2\left(\overline{s_i s_j}\right)_{vacuum} = dA_i \int_{\omega_i=0}^{\omega_i=2\pi} \frac{\cos\theta_i d\omega_i}{\pi} = dA_i = d_i\Sigma\left(\overline{s_i s_j}\right)_{vacuum}$$

By integrating the final equation over A_i:

$$A_i = \sum_{j=1}^{n}\left(\overline{s_i s_j}\right)_{vacuum}$$

Now, considering the container to be filled with a gas, an average proportion α_g of each ray moving from P_i to P_j will be absorbed by the gas, and an average proportion τ_g will in all directions, be transmitted to the walls "seen" by P_i. Evidently:

$$\alpha_{gi} = \tau_{gi} = 1$$

Hence:

$$A_I = \alpha_{gi}\sum_{j=1}^{n}\left(\overline{s_i s_j}\right)_{vacuum} + \tau_{gi}\sum_{j=1}^{n}\left(\overline{s_i s_j}\right)_{vacuum}$$

Take:

$$\overline{s_i g} = \alpha_{gi} \sum_{j=1}^{n} \left(\overline{s_i s_j}\right)_{vacuum} \text{ and } \overline{s_i s_j} = \tau_{gi} \left(\overline{s_i s_j}\right)_{vacuum}$$

We obtain:

$$\overline{s_i g} + \sum_{j=1}^{n} \overline{s_i s_j} = A_i$$

NOTE.–

Consider a gray gas held in a container with a black wall and a total surface A_T. The energy flow emitted by the gas to the wall element A_i is known:

$$A_i \varepsilon_g E_g = E_g \overline{g s_i}$$

Thus:

$$\overline{g s_i} = \varepsilon_g A_i$$

If the black surface A_i emits a flow $A_T E_N$ toward the gas, the gas will absorb:

$$\alpha_{gi} A_i E_{Ni} = E_{Ni} \overline{s_i g}$$

Hence:

$$\overline{s_i g} = \alpha_{gi} A_i$$

We have not considered temperatures T_G and T_i to be equal. In this case:

$$\alpha_{gi} \neq \varepsilon_g \text{ ; hence : } \overline{s_i g} \neq \overline{g s_i}$$

Moreover, by definition:

$$\overline{s_i s_j} = \tau_{ij}\left(s_i s_j\right)_{\text{vacuum}} = \left(1-\alpha_{ij}\right)\left(s_i s_j\right)_{\text{vacuum}}$$

And reciprocally:

$$\overline{s_j s_i} = \left(1-\alpha_{ji}\right)\left(\overline{s_j s_i}\right)_{\text{vacuum}}$$

Therefore:

$$s_j s_i \neq s_i s_j \text{ since } \alpha_{ij} \neq \alpha_{ji} \text{ if } T_i \neq T_j$$

Note that between equivalent direct transfer surfaces in a vacuum, which are simply geometric parameters:

$$\left(\overline{s_j s_i}\right)_{\text{vacuum}} = \left(\overline{s_i s_j}\right)_{\text{vacuum}}$$

While the matrix of $\left(\overline{s_i s_j}\right)_{\text{vacuum}}$, which is purely geometric in nature, is symmetrical, the same is not true for the matrix of $\left(\overline{s_i s_j}\right)$.

1.3.11. *Energy flow density emitted by each wall element*

The total flow emitted by the surface A_i is distributed as follows:

$$\varphi_{ei} A_i = \varepsilon_i E_i A_i + \rho_i \lambda_{Ii} A_i \qquad [1.7]$$

E_i: black emittance corresponding to temperature T_i of A_i

ρ_i: reflectivity of surface A_i

φ_{Ii}: incident flow density on surface $A_{i.}$

The incident flow on A_i comes from:

– the gas: $E_g\ \overline{gs_i}$

– all surfaces A_i, including surface A_i insofar as the surface "sees" itself, i.e.:

$$\sum_{j=1}^{n} \varphi_{ej} \overline{s_j s_i}$$

Therefore:

$$\varphi_{Ii} A_i = E_g \overline{gs_i} + \sum_{j=1}^{n} \varphi_{ej} \overline{s_j s_i} \qquad [1.8]$$

By eliminating $\phi_{Ii}A_i$ in equations [1.7] and [1.8], we obtain a linear equation in terms of φ_{ej}:

$$\sum_{j=1}^{n} \left(\overline{s_j s_i} - \delta_{ij} \frac{A_j}{\rho_j} \right) \varphi_{ej} = -\frac{A_i \varepsilon_i E_i}{\rho_i} - E_g \overline{gs_i}$$

There are n equations of this type, i.e. as many equations as the number of surface elements A_i (and unknown quantities φ_{ei}). δ_{ij} is the Kronecker symbol (zero if $i \neq j$ and equal to 1 if $i = j$). Note that the equivalent direct transfer surface $\overline{s_j s_i}$ relates to radiation emanating from A_j and received by A_i (see section 1.3.7).

This system of n equations is the "characteristic system" of the heat transfer, and may be written in matrix form as follows (where, unusually, the first index in each term characterizes the column and not the line):

$$\begin{bmatrix} \overline{s_1 s_1} - \frac{A_1}{\rho_1} & \overline{s_2 s_1} & \overline{s_3 s_1} & s_j s_1 \\ \overline{s_1 s_2} & \overline{s_2 s_2} - \frac{A_2}{\rho_2} & \overline{s_3 s_2} & s_j s_2 \\ \overline{s_1 s_3} & \overline{s_2 s_3} & \overline{s_3 s_3} - \frac{A_3}{\rho_3} & s_j s_3 \\ \overline{s_1 s_i} & \overline{s_2 s_i} & & s_j s_i \end{bmatrix} \cdot \begin{bmatrix} \varphi_{e1} \\ \varphi_{e2} \\ \varphi_{e3} \\ \varphi_{ej} \end{bmatrix} = \begin{bmatrix} -\frac{A_1 \varepsilon_1 E_1}{\rho_1} - E_g \overline{gs_1} \\ -\frac{A_2 \varepsilon_2 E_2}{\rho_2} - E_g \overline{gs_2} \\ -\frac{A_3 \varepsilon_3 E_3}{\rho_3} - E_g \overline{gs_3} \\ -\frac{A_i \varepsilon_i E_i}{\rho_i} - E_g \overline{gs_i} \end{bmatrix}$$

Or, in condensed form:

$$(D)(\varphi_e)=(\xi)$$

As we have seen, the transmissivity is dependent on the direction of propagation. Thus, in the matrix shown above, the first index relates to the emitting surface element (the source) and the second relates to the receptor surface (target), which is unique for each line.

The *column vector* of the second member includes the emissive power of the walls and of the gas, evaluated at the corresponding temperatures. This vector shows the thermal state of the system, and in this case is noted (ξ). This vector has also been referred to as the excitation vector by certain authors.

The determinant D which characterizes this system of linear equations may be developed in relation to the jth column:

$$D=\sum_{i=1}^{n} d_{ij}\left((-1)^{1+j} M_{ij}\right)=\sum_{i=1}^{n} d_{ij}D_{ij}$$

(M_{ij} is the minor determinant relating to term d_{ij} of the matrix (D) and is obtained by eliminating line i and column j from D. The determinant D_{ij} is the "cofactor" of term d_{ij}).

Solutions to a linear system of this type are obtained by replacing D in column j by the vector (ξ) with components ξ_i. The determinant D_j obtained in this way is written:

$$D_j=\sum_{i=1}^{n}\xi_i D_{ij}$$

The energy flow density emitted by element j is:

$$\varphi_{ej}=\frac{D_j}{D}$$

1.3.12. *Net flow received by surface Ai*

This energy level determines the accumulation of heat in the wall A_i, and is particularly useful in practice.

By definition, the net received flow is:

$$\phi_{net.i} = A_i \varphi_{net.i} = A_i \left(\varphi_{li} - \varphi_{ei} \right) \qquad [1.9]$$

Evidently, the net received flow is the difference between the incident flow and the emitted flow. However, we know that, considering these flow densities in relation to A_i:

$$\varphi_{ei} = \varepsilon_i E_i + \rho_i \varphi_{li} \qquad [1.10]$$

By eliminating φ_{li} in [1.9] and [1.10]:

$$\varphi_{net.i} = \frac{\varepsilon_i}{\rho_i}\left(\varphi_{ei} - E_i \right) \text{ or, alternatively: } \phi_{net.i} = \frac{\varepsilon_i A_i}{\rho_i}\left(\varphi_{ei} - E_i \right)$$

1.3.13. *Difference between net received flow and absorbed flow*

We know that the net flow received by an element of surface A is:

$$\phi_{net} = \phi_l - \phi_e$$

By definition, the flow absorbed by surface A is:

$$\phi_a = \alpha_A \phi_I$$

However, the emitted flow is:

$$\phi_e = \underset{\text{source term}}{\varepsilon EA} + \underset{\text{reflection term}}{\rho A \varphi_I} = \varepsilon EA + \left(1 - \alpha_A\right)\phi_I$$

Hence:

$$\phi_{net} = \phi_l - \phi_a = -\varepsilon EA < 0$$

This means that the wall loses energy (and thus cools) by radiation alone, and this energy must be deducted from the absorbed energy.

1.3.14. *Net flow for a gray surface surrounding a gray gas*

The flow density emitted by surface A is distributed as follows (in terms of density):

$$\varphi_e = \varepsilon_A E_A + \rho_A \varphi_I$$

Following Kirchhoff's law, for the wall:

$$\rho_A = 1 - \alpha_A = 1 - \varepsilon_A$$

The origin of the incident flow is:

$$\varphi_J = \tau_g \varphi_e + \varepsilon_g E_g$$

However, the transmissivity from one point A to the other through the gas is:

$$\tau_g = 1 - \alpha_g$$

$$\varphi_e = \varepsilon_A E_A + (1 - \varepsilon_A)\varphi_I$$

$$\varphi_I = (1 - \alpha_g)\varphi_e + \varepsilon_g E_g$$

By eliminating ϕ_I:

$$\varphi_e = \frac{\varepsilon_A E_A + (1 - \varepsilon_A)\varepsilon_g E_g}{\alpha_g + \varepsilon_A - \alpha_g \varepsilon_A}$$

The net received flow density is:

$$\varphi_{net} = \frac{\varepsilon_A}{1 - \varepsilon_A}(\varphi_e - E_A) \qquad [1.11]$$

$$\varphi_{net} = \frac{\varepsilon_A \varepsilon_g E_g - \varepsilon_A \alpha_g E_A}{\alpha_g + \varepsilon_A - \alpha_g \varepsilon_A}$$

1.3.15. *Net flow for two surfaces in close proximity*

In this case, the thickness of the gas between the two surfaces is sufficiently low for the gas to be considered totally transparent:

$$\tau_g = 1 \qquad \alpha_g = 0$$

The distribution of flow densities emitted by each surface is given by the following equations:

$$\varphi_{e1} = \varepsilon_1 E_1 + (1-\varepsilon_1)\varphi_{I1}$$

$$\varphi_{e2} = \varepsilon_2 E_2 + (1-\varepsilon_2)\varphi_{I2}$$

The origin of the incident flows is such that:

$$\varphi_{I1} = \varphi_{e2} \quad \text{and} \quad \varphi_{I2} = \varphi_{e1}$$

By eliminating φ_{I1}, φ_{I2} and φ_{e2} from these four equations, the flow emitted by surface 1 is:

$$\varphi_{e1} = \frac{\varepsilon_1 E_1 + (1-\varepsilon_1)\varepsilon_2 E_2}{1-(1-\varepsilon_1)(1-\varepsilon_2)}$$

Hence, the net flow density received by surface 1 using equation [1.11] is:

$$\varphi_{net.1} = \frac{\varepsilon_1}{1-\varepsilon_1}(\varphi_{e1} - E_1) = \frac{E_2 - E_1}{\dfrac{1}{\varepsilon_1} + \dfrac{1}{\varepsilon_2} - 1}$$

Note the antisymmetric nature of this expression.

1.3.16. *Evaluating EDTS. Some simple specific examples*

1) EDTS for self-transfer:

If the whole of an element of surface A_i can be "seen" from any point on this surface, i.e. if A_i is totally concave, then:

$$\left(\overline{s_i s_i}\right)_{vacuum} = A_i \text{ and } \overline{s_i s_i} = \tau_{gii} A_i$$

The transmissivity is evaluated using the average value for the distance between any given pair of points on the surface A_i.

If A_i is flat or completely convex on the inside, then:

$$\left(\overline{s_i s_i}\right)_{vacuum} = 0 \text{ and } \overline{s_i s_i} = 0$$

2) EDTS for cross-transfer:

The EDTS between A_p and A_p in a vacuum can be shown to be the sum of the crossed lines between the extremities of A_p and A_P, minus the sum of the uncrossed lines between the extremities, and finally divided by 2:

$$\left(\overline{s_p s_P}\right)_{vacuum} = \frac{1}{2}\left(\left(P_2 p_1 + P_1 p_2\right) - \left(P_1 p_1 + P_2 p_2\right)\right)$$

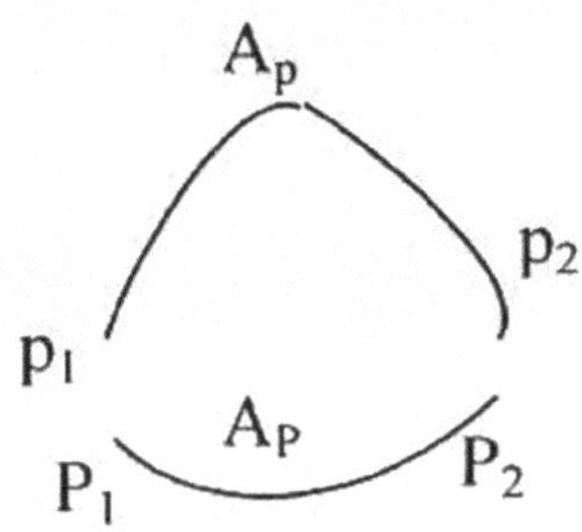

Figure 1.7. *EDTS calculation*

When p_1 and P_1 are the same and p_2 and P_2 are also the same, then:

$$\left(\overline{s_p s_P}\right)_{vacuum} = \frac{1}{2}\left(p_1 p_2 + P_1 P_2\right) = p_1 p_2 = P_1 P_2 = \left(\overline{s_P s_p}\right)_{vacuum}$$

While EDTS in a vacuum are symmetrical, this ceases to be true in the presence of gas, as the transmissivity depends on the temperature of the emitting surface:

$$\overline{s_p s_P} \neq \overline{s_P s_p}$$

NOTE.–

De Vriendt [DE 84] and Hottel *et al.* [HOT 67] provide more detailed information regarding the calculation of EDTS.

1.4. Heat transfer in rotary calcination kilns

1.4.1. *Transfer mechanism*

In this case, the aim is to transfer heat from the hot gas to the product requiring calcination. The cross-section of a kiln is shown in Figure 1.8:

1) The gas *loses* heat by radiation and convection:

- to the free surface of the product (power W_{GPg});

- to the part of the wall not covered by the product (power W_{Gpg}).

2) The product *receives* heat:

- by radiation and convection from the gas over its free surface (power W_{GPg});

- by radiation from the uncovered section of wall, through the gas and over the free surface of the product (power W_{pPr});

- by conduction and radiation from the wall with which it is in direct contact (power W_{pPc}).

3) The wall:

- *receives* heat from the gas by radiation and convection (power W_{Gpg});

- *returns* this heat to the product, either by direct radiation (power W_{pPc}) onto the free surface of the product, or by radiation and conduction from the surface covered by the product (power W_{pPr}).

The overall heat lost by the gas and recuperated by the product may be written as:

$$W_{GPg} + W_{Gpg} = W_{GPg} + W_{pPr} + W_{pPc}$$

i.e.:

$$W_{Gpg} = W_{pPr} + W_{pPc}$$

This equality simply shows the energy balance of the wall acting as a regenerator, i.e. accumulating heat received by radiation and returning it to the product in the same manner.

1.4.2. *Equivalent direct transfer surface for the bare wall and free surface of the product*

Let p and P represent the bare wall (i.e. the area not covered by the product) and the free surface of the product, respectively, and let A_p and A_P be the corresponding surfaces.

Let θ be the fraction of the circumference of the shell covered by the product:

$$\theta = \frac{t_F}{t_F + t_C} \qquad [1.12]$$

t_F: time spent being cooled by product for any point on the wall

t_C: time spent being heated by gas for any point on the wall.

The period of rotation is, evidently:

$$T = t_F + t_C$$

If D is the diameter of the kiln:

$$A_p = \pi D(1-\theta) \text{ and } A_P = D\sin\frac{0}{2} \qquad [1.13]$$

For the container holding the body of gas:

$$\overline{s_p s_p} = \tau_{pp} A_p \qquad \overline{s_P s_P} = 0 \quad (\text{flat surface})$$

We also know that:

$$\left(\overline{s_p s_P}\right)_{\text{vacuum}} = \left(\overline{s_P s_p}\right)_{\text{vacuum}} = P_1 P_2 = A_P$$

Hence:

$$\overline{s_p s_P} = \tau_{pP} A_P \quad \text{and} \quad \overline{s_P s_p} = \tau_{Pp} A_p$$

$$\overline{g s_P} = \varepsilon_g A_P \qquad \overline{g s_p} = \varepsilon_g A_p$$

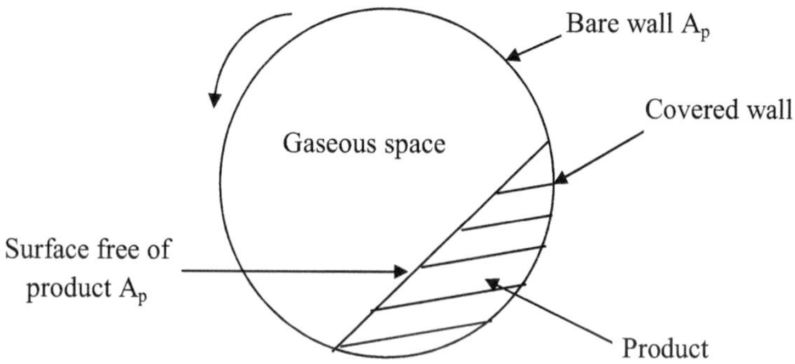

Figure 1.8. *Rotary calcination kiln*

We know that, apart from the temperature of the gas T_G, τ_{ij} is solely dependent on the temperature of the emitting surface and on the average distance traveled L_{ij} between the emitting and receiving points.

For a rotating kiln, we may consider that:

$$L_{PP} = L_{pP} = L_{Pp} = 0.94D$$

D: internal diameter of the kiln (m)

Hence, for the transmissivities, we have:

$$\tau_{PP} = \tau_{pP} = \tau_p = f\left(T_P,\ T_G D\right)$$

$$\tau_{Pp} = \tau_p = g\left(T_P,\ T_G D\right)$$

Finally:

$$\overline{s_p s_p} = \tau_p A_p \qquad \overline{s_P s_P} = 0$$

$$\overline{s_p s_P} = \tau_p A_P \qquad \overline{s_P s_p} = \tau_P A_P$$

$$\overline{g s_p} = \varepsilon_g A_p \qquad \overline{g s_P} = \varepsilon_g A_P$$

1.4.3. *Net flow densities received by the wall and the product*

Given that the index p characterizes the surface of the refractory wall and the index P characterizes the free (flat) surface of the product, the typical radiation system is written as:

$$\begin{bmatrix} \overline{s_p s_p} - \dfrac{A_p}{\rho_p} & \overline{s_P s_p} \\ \overline{s_p s_P} & \overline{s_P s_P} - \dfrac{A_P}{\rho_P} \end{bmatrix} \times \begin{bmatrix} \varphi_{ep} \\ \varphi_{eP} \end{bmatrix} = \begin{bmatrix} -\dfrac{A_p \varepsilon_p E_p}{\rho_p} & - & E_g \overline{g s_p} \\ -\dfrac{A_P \varepsilon_P E_P}{\rho_P} & - & E_g \overline{g s_P} \end{bmatrix}$$

i.e.:

$$\begin{bmatrix} \tau_p A_p - \dfrac{A_p}{\rho_p} & \tau_P A_P \\ \tau_p A_P & -\dfrac{A_P}{\rho_P} \end{bmatrix} \times \begin{bmatrix} \varphi_{ep} \\ \varphi_{eP} \end{bmatrix} = \begin{bmatrix} -\dfrac{A_p \varepsilon_p E_p}{\rho_p} & - & E_g \varepsilon_g A_p \\ -\dfrac{A_P \varepsilon_P E_P}{\rho_P} & - & E_g \varepsilon_g A_p \end{bmatrix}$$

Take:

$$[D] \times [\varphi_e] = [\xi]$$

With a system determinant:

$$D = A_P A_p \left(\frac{1}{\rho_P \rho_p} - \frac{\tau_p}{\rho_P} - \tau_p \tau_P \right)$$

And:

$$D_p = \begin{bmatrix} \tau_p A_p - \dfrac{A_p}{\rho_p} & -\dfrac{A_p \varepsilon_p E_p}{\rho_p} - E_g \varepsilon_g A_p \\ \tau_p A_p & -\dfrac{A_P \varepsilon_P E_P}{\rho_P} - \varepsilon_g E_g A_P \end{bmatrix}$$

$$D_P = \frac{A_p A_P}{\rho_p} \left(\varepsilon_g E_g + \varepsilon_p E_p \tau_p + \frac{\varepsilon_P E_P \left(1 - \tau_p \rho_p\right)}{\rho_P} P \right)$$

Take:

$$\pi = \frac{A_P}{A_p} \qquad [1.14]$$

Hence, for the net flow density received by the product:

$$\phi_{net.P} = \varepsilon_P \frac{\varepsilon_g E_g + \varepsilon_p E_p \tau_p - E_P \left(1 - \tau_p \rho_p - \pi \tau_p \tau_P \rho_p\right)}{1 - \rho_p \tau_p - \pi \tau_p \tau_P \rho_p \rho_P} \qquad [1.15]$$

$$D_p = \begin{bmatrix} -\dfrac{A_p \varepsilon_p E_p}{\rho_p} - \varepsilon_g E_g A_p & \tau_P A_P \\ -\dfrac{\varepsilon_P E_P A_P}{\rho_P} - \varepsilon_g E_g A_P & -\dfrac{A_P}{\rho_P} \end{bmatrix}$$

$$D_p = \frac{A_p A_P}{\rho_p} \left(\varepsilon_g E_g \left(1 + \tau_p\right) + \varepsilon_p E_p \pi \tau_p + \frac{\varepsilon_p E_p}{p_P} \right)$$

$$\phi_{net.P} = \varepsilon_P \frac{\varepsilon_g E_g \left(1 + \pi \rho_P \tau_P\right) + \varepsilon_P E_P \pi \tau_P - E_P \left(1 - \tau_p - \pi \tau_p \tau_P \rho_P\right)}{1 - \rho_p \tau_p - \dfrac{A_P}{A_p} \tau_p \tau_P \rho_p \rho_P}$$

Take:

$$a_g = \frac{\varepsilon_g\left(1+\pi\rho_P\tau_P\right)}{1-\tau_p-\pi\tau_p\tau_P\rho_P} \quad \text{and} \quad a_P = \frac{\varepsilon_P\tau_P\pi}{1-\tau_p-\pi\tau_p\tau_P\rho_P} \qquad [1.16]$$

$$R_C^{'} = \varepsilon_p \frac{1-\tau_p-\pi\tau_p\tau_P\rho_P}{1-\rho_p\tau_p-\pi\tau_p\tau_p\rho_p\rho_P}\sigma \qquad [1.17]$$

$$E_g^{'} = \sigma\left(T_G^{'}\right)^4 = a_g E_g + a_P E_P \qquad [1.18]$$

This equation enables us to determine $T_G^{'}$.

The net flow density received by the wall is thus:

$$\left(\varphi_{net.p}\right)_0 = R_C^{'}\left(\left(T_G^{'}\right)^4 - \left(T_p\right)^4\right)$$

The zero index indicates that we have not yet considered the additional chemical heat transfer which occurs due to convection.

1.4.4. *Regeneration from the wall*

The temperature in a kiln is approximately 1,500°C. The metal wall must therefore be protected by a refractory coating, with a thickness of approximately 15 cm. Not all of the thickness of this insulating layer is involved in regeneration, as its external surface remains close to the ambient temperature; the truly active part of the lining e_a has a thickness of approximately 0.05 m.

Let us consider a rectangular unit of space on the wall of the rotating cylinder, with one side length corresponding to the unit of length of the perimeter of the shell, and the other to the unit of length of the axis.

The active unitary volume will have a thermal capacity of:

$$e_a\rho_p C_p$$

ρ_p: bulk density of the wall ($kg.m^{-3}$)

C_p: bulk thermal capacity of the wall ($J.\ kg^{-1}{}^{\circ}C^{-1}$)

e_a: active thickness (m).

When the wall is in a state of thermal equilibrium, we can write:

$$e_a \rho_o C_p \frac{dT_p}{dt} = 0$$

T_p: average temperature of the active surface (K)

t: time (s).

When considering convection or conduction transfer, characterized by the additional coefficient α_C, in addition to radiation, we must write, for example, in the case of heating by a gas to T_G:

$$\varphi_{net.p} = \left(\varphi_{net.p}\right)_o + \alpha_C \left(T_G - T_p\right)$$

Also:

$$T_G - T_p = \frac{T_G^4 - T_p^4}{4T_m^3} \text{ where } T_m \# \frac{T_G + \sqrt{T_G T_p}}{2} \qquad [1.19]$$

And:

$$\left(\varphi_{net.p}\right)_o = R_C\left(\left(T_G\right)^4 - T_P^4\right)$$

Hence:

$$\varphi_{net.p} = R_C\left(\left(T_G\right)^4 - T_P^4\right) + \frac{\alpha_C}{4T_m^3}\left(T_G^4 - T_P^4\right)$$

Taking:

$$\left(T_G^*\right)^4 = \frac{R_C\left(T_G^{'}\right)^4 + \left(\alpha_C/4T_m^3\right)T_G^4}{R_C + \left(\alpha_C/4T_m^3\right)} \qquad [1.20]$$

We obtain:

$$\varphi_{net.p} = \left(R'_C + \alpha_C/4T_m^3\right)\left(\left(T_G^*\right)^4 - T_p^4\right) = R_C^*\left(\left(T_G^*\right) - T_p^4\right) \quad [1.21]$$

In the case of cooling by radiation and conduction from a product at temperature T_P, we obtain:

$$\varphi_{net.p} = \left(R_F + \alpha_C/4T_m^3\right)\left(T_p^4\left(t_p^*\right)^4\right) = R_F^*\left(T_p^4 - \left(t_p^*\right)^4\right)$$

$$R_F^* = R_F + \frac{\alpha_C}{4\left(T_m\right)^4} \qquad t_p^* = t_p \qquad T_m \# \frac{t_p + \sqrt{t_p T_G}}{2}$$

1.4.5. *Expression of additional coefficients for heat transfer*

1) Gas convection:

$$\alpha_C = 23.6\ \phi_{GM}^{0.67}$$

α_C: convection coefficient (W.m^2.°C)

φ_{GM}: mass flow density of the gas (applied to the section of area through which the gas is able to pass) (kg.m^{-2}.s^{-1})

2) Coefficient representing heat transfer through direct contact between the product and the wall (based on experimental results obtained by the author):

$$\alpha_C = 80\ \text{W.m}^{-2}.°\text{C}^{-1}$$

1.4.6. *Solution to the wall equation*

As the shell turns, the wall is subject to a periodic thermal cycle, which can be resumed as follows:

$$\text{cooling} - \rightarrow t_0 - \rightarrow \text{heating} - \rightarrow T_0 - \rightarrow \text{cooling} - \rightarrow t_0 \ \text{etc.}$$

The wall equation is a classic differential equation with separated variables, and a solution has been published by De Vriendt [DE 84, p. 204], among others.

The wall heating and cooling phases will now be examined separately.

1) Heating of the wall:

$$\frac{dT_p}{dt} = \frac{R_C^*}{e_a \rho_p C_p}\left(\left(T_G^*\right)^4 - T_p^4\right)$$

Using the identity,

$$\frac{1}{T_G^4 - T_p^4} = \frac{1}{2T_G^2}\left(\frac{1}{T_G^2 - T_p^2} + \frac{1}{T_G^2 + T_p^2}\right)$$

The solution is

$$\frac{1}{2}\mathrm{Ln}\left[\left(\frac{T_G - t_0}{T_G + t_0}\right)\times\left(\frac{T_G + T_0}{T_G - T_0}\right)\right] + \mathrm{Arc\ tg}\frac{T_0}{T_G} - \mathrm{Arc\ tg}\frac{t_0}{T_G} = \frac{2T_G^3 R_C^* t_C}{\rho_p C_p e_a} = K_C$$

The reciprocal of the quantity under the logarithm may be written as:

$$1 - \frac{2T_G\left(T_0 - t_0\right)}{T_G^2 - T_0 t_0 + T_0\left(T_0 - t_0\right)} = 1 - \frac{2T_G \Delta T}{T_G^2 - T_0 t_0 + T_0 \Delta T}$$

However,

$$\frac{1}{2}\mathrm{Ln}\frac{1}{1-a} \# \frac{1}{2}a$$

Additionally,

$$\mathrm{Arc\ tg}\frac{T_0}{T_G} - \mathrm{Arc\ tg}\frac{t_0}{T_0} = \mathrm{Arc\ tg}\frac{T_0 \Delta T}{T_G^2 + T_0 t_0}$$

Taking the first-order series expansion:

$$T_G \Delta T\left(\frac{1}{T_G^2 - T_0 t_0 + T_G \Delta T} + \frac{1}{T_G^2 + T_0 t_0}\right) = \frac{2T_G^{*3} R_C^* \tau_C}{\rho_p C_p e_a} = K_C \qquad [1.22]$$

2) Cooling of the wall:

Using the same approach, we obtain:

$$t_p \Delta T \left(\frac{1}{T_0 t_0 - t_p^2 + t_p \Delta T} - \frac{1}{t_p^2 + T_0 t_0} \right) = \frac{2T_p^3 R_F^* \tau_F}{\rho_p C_p e_a} = K_F \quad [1.23]$$

In these expressions, τ_C and τ_F are the heating and cooling times for the wall, and if T is the rotation period of the shell, then:

$$T = \tau_C + \tau_F$$

Let us use the following unknown quantities:

$$\Delta T = T_0 - t_0 \text{ and } \tau^2 = T_0 t_0$$

In both of the previous equations, we can *ignore the ΔT element of the denominator*, and the equations then become easy to solve:

$$\tau^4 = \frac{K_C T_G + K_F t_p}{\frac{K_C}{T_G^3} + \frac{K_F}{t_p^3}} \text{ and } \Delta T = \frac{T_G^4 - t_p^4}{2\left(\frac{t_p^3}{K_F} + \frac{T_G^3}{K_C} \right)} \quad [1.24]$$

However:

$$\text{Since } T_0 = t_0 + \Delta T, \text{ we have: } \tau^2 = (t_0 + \Delta T) t_0$$

The solution to this second-degree equation gives:

$$\overline{T_p} = \left[\tau^2 + \left(\frac{\Delta T}{2} \right)^2 \right]^{1/2} \quad [1.25]$$

$$T_0 = \overline{T_p} + \frac{\Delta T}{2} \quad \text{and} \quad t_0 = \overline{T_p} - \frac{\Delta T}{2}$$

NOTE.–

Coefficients K_F and K_C are inversely proportional to the active thickness e_a of the wall. Furthermore, we know that $\Delta T/T_m$ is considerably smaller than 1, and therefore, in reality, the two differential equations for the wall give us a product eaΔT and not an absolute value ΔT.

The regeneration process is defined by the heat which is accumulated then re-transmitted to the product, a value proportional to $e_a\Delta T$.

The possibility of error is greatly reduced if we simply *estimate* the active thickness to be 0.05 m, as above.

1.4.7. *Overall thermal power transferred to the product for calcination*

Our reasoning is based on a length of one unit along the axis of the shell.

The product receives heat in two ways:

The free surface A_P of the product is in contact with the gas in the system, and it receives the following quantity of thermal power:

$$A_P \varphi_{net,\,P,\,r}$$

The product also receives thermal power by a regeneration effect at points of contact with the wall:

$$A_{PC}\varphi_{net,\,P,\,C} = -A_{PC}\varphi_{net,\,p,\,C}$$

Giving a total of:

$$\varphi_{net,\,P} = A_P \times \varphi_{net,\,P,1} + A_{PC} \times \varphi_{net,\,P,\,C}$$

This thermal power input enables the calcination reaction to occur and may also result in heating of the product.

EXAMPLE 1.6.–

$t_F = 10$ s $\quad T_P = 800°C = 1{,}073$ K $\quad D = 2.3$ m

$t_C = 20$ s $\quad T_G = 1{,}500°C = 1{,}773$ K $\quad e_a = 0.05$ m

$\rho_p = 2{,}000$ kg.m^{-3} $\quad \phi_{GM} = 0.32$ kg.s^{-1}.m^{-2} $\quad C_p = 900$ J.kg^{-1}.°C^{-1}

$P_c = 0.15$ atm $\quad P_{totale} = 1$ atm $\quad P_v = 0.10$ atm

$\varepsilon_P = 0.8$ $\quad \varepsilon_p = 0.8$

$$L_m = 0.94 \times 0.23 = 2.16 \text{ m} \quad [1.26]$$

The angle at the center subtended by the solid is:

$$360 \times \frac{10}{10+20} = 120° \quad [1.27]$$

$$A_P = 2 \times \frac{2.3}{2} \times \sin\frac{120}{2} = 1.992 \text{ m\#2m} \quad [1.28]$$

$$A_p = \frac{20}{10+20} \times \pi \times 2.3 = 4.817 \text{ m}$$

$$\pi = \frac{A_P}{A_p} = 1.992 / 4.817 = 0.413 \quad [1.29]$$

Let us calculate the emissivity of the gas at 1,500°C (Figures 1.7 and 1.8):

$P_cL_m = 0.15 \times 2.16 = 0.33$ m.atm $\quad P_vL_m = 0.10 \times 2.16 = 0.22$ m.atm

$\varepsilon_c = 0.11$ $\quad \varepsilon_v = 0.11$

$$\varepsilon_g = 0.11 + 0.11 - 0.03 = 0.19 \quad [1.30]$$

Let us calculate the absorptivity of the gas at 1,500°C in contact with the product at 800°C:

At 800°C:

$$P^c L_m \frac{1073}{1773} = 0.20 \text{ m.atm} \qquad P_v L_m \times \frac{1073}{1773} = 0.13 \text{ m.atm} \qquad [1.31]$$

$$\varepsilon_c = 0.132 \qquad \varepsilon_v = 0.16$$

$$\alpha_{gP} = 0.132\left(\frac{1773}{1073}\right)^{0.65} + 0.16\left(\frac{1773}{1073}\right)^{0.45} - 0.03 \qquad [1.32]$$

$$= 0.183 + 0.200 - 0.03$$

$$\alpha_{gP} = 0.353$$

The fraction of the radiation coming from the product and transmitted by the gas is:

$$\tau_P = 1 - 0.353 = 0.647$$

Absorptivity of the gas at 1,500°C in contact with the wall at $\overline{T_P}^{(0)}$:

$$\overline{T_P}^{(0)} = \sqrt{1073 \times 1773} = 1380 \text{ K}$$

$$P_c L_m \times \frac{1380}{1773} = 0.26 \text{ m/atm} \qquad P_v L_m \times \frac{1380}{1773} = 0.17 \text{ m/atm} \qquad [1.33]$$

$$\varepsilon_c = 0.105 \qquad \varepsilon_v = 0.11$$

$$\alpha_{gp} = 0.105\left(\frac{1773}{1380}\right)^{0.65} + 0.11\left(\frac{1773}{1380}\right)^{0.45} - 0.03 \qquad [1.34]$$

$$= 0.123 + 0.123 - 0.03$$

$$\alpha_{gp} = 0.216$$

The fraction of the radiation coming from the wall and transmitted by the gas is:

$$\tau_P = 1 - 0.16 = 0.784 \quad [1.35]$$

$$a_g = \frac{0.19(1 + 0.413 \times 0.647 \times 0.2)}{1 - 0.784 - 0.413 \times 0.647 \times 0.784 \times 0.2} = 1.149 \quad [1.36]$$

$$a_P = \frac{0.8 \times 0.784 \times 0.413}{0.174} = 1.4887 \quad [1.37]$$

$$\left(\frac{T_G'}{100}\right)^4 = 1.149 \times (17.73)^4 + 1.4887(10.73)^4 = 133724 \quad [1.38]$$

$$T_G' = 1910 \text{ K}$$

$$R_C' = \frac{5.67 \times 10^{-8} \times (1 - 0.784 - 0.413 \times 0.784 \times 0.647 \times 0.2)0.8}{1 - 0.784 \times 0.2 - 0.413 \times 0.647 \times 0.784 \times 0.2 \times 0.2} \quad [1.39]$$

$$R_C = 0.9454 \times 10^{-8}$$

$$\alpha_C = 23.6 \times (0.32)^{0.67} = 11 \text{ W.m}^{-2}\,°\text{C}^{-1} \quad \text{(see 1.4.5)}$$

$$T_m = \frac{1773 + \sqrt{1073 \times 1773}}{2} = 1576 \text{ K} \quad [1.40]$$

$$R_C^* = \left(0.9454 \times 10^{-8} + \frac{11}{(1573)3 \times 4}\right) = 1.016 \times 10^{-8} \quad [1.41]$$

$$\left(\frac{T_G'}{100}\right) = 1.149 \times 17.73^4 + 1.4887 \times 10.73^4$$

$$T_G' = 1894 \text{ K}$$

$$\left(\frac{T_G^*}{100}\right)^4 = \frac{1.016 \times 10^{-8} \times 18.94^4 + \left[11/\left(4 \times 1576^3\right)\right]17.73^4}{1.016 \times 10^{-8} + 0.07025 \times 10^{-8}} \quad [1.42]$$

$$T_G^* = 1886 \text{ K}$$

Let us calculate the parameters of the net flow lost by the shell due to cooling on contact with the product:

$$-\phi_{net.p} = \frac{5.67\times10^{-8}\left(T_P^4 - T_p^4\right)}{\frac{1}{0.8}+\frac{1}{0.8}-1} = R_F\left(T_P^4 - T_p^4\right) \qquad \text{(see section 1.3.15)}$$

$$R_F = 3.78\times10^{-8}$$

$$T_m = \frac{1073+\sqrt{1073\times1773}}{2} = 1226\ \text{K}$$

$$R_F^* = 3.78\times10^{-8} + \frac{80}{4\times(12.26)^4} = 3.81\times10^{-8} \qquad \text{(see section 1.4.4)}$$

Now, calculate the maximum temperature T_0 and minimum temperature t_0 of the wall:

$$K_C = \frac{1.016\times10^{-8}\times2\times(1886)^3\times20}{2000\times900\times0.05} = 0.0303 \qquad [1.43]$$

$$K_F = \frac{3.81.10^{-8}\times2\times(1073)^3\times10}{2000\times900\times0.05} = 0.0104 \qquad [1.44]$$

$$\Delta T = \frac{1905^4 - 1073^4}{2\left(\frac{1073^3}{0.0104}+\frac{1905^3}{0.030}\right)} = 16.5°\text{C} \qquad [1.45]$$

$$\tau^4 = \frac{0.030\times1886+0.0104\times1073}{\frac{0.030}{6.7085\times10^9}+\frac{0.0104}{1.2354\times10^9}} = 5.2551\times10^{12} \qquad [1.46]$$

$$\tau^2 = 2292412$$

$$\overline{T_p} = (2292412+68)^{1/2} = 1514\ \text{K} \qquad [1.47]$$

$$T_0 = 1522\ \text{K} = 1249°\text{C} \qquad t_0 = 1506\ \text{K} = 1233°\text{C}$$

Net flow density over the free surface of the product (by radiation and convection):

$$\varphi_{net.P.r} = 0.8 \times 5.67 \times \frac{0.19 \times (17.73)^4 + 0.8(15.14)^4 \times 0.784 - (10.73)^4 (1 - 0.784 \times 0.2 - 0.,413 \times 0.784 \times 0.647 \times 0.2)}{1 - 0.2 \times 0.784 - 0.413 \times 0.647 \times 0.784 \times 0.2 \times 0.2}$$

$$\varphi_{net,\,P,\,r} = 223366 \text{ W.m}^{-2} \qquad [1.48]$$

Net flow per meter along kiln axis:

$$\phi_{net,\,P,\,r} = 223366 \times 2 = 446732 \text{ W.m}^{-1}$$

Net flow density by direct contact (regeneration effect):

$$\varphi_{net,\,P,\,c} = -\varphi_{net,\,p,\,c} = 3.81\left(15.14^4 - 10.73^4\right) = 149967 \text{ W.m}^{-2}$$

Net flow per meter along kiln axis:

$$\phi_{net,\,P,\,c} = 149967\left(\frac{10}{30} \times \pi \times 2.3\right) = 361167 \text{ W.m}^{-1}$$

That gives a total, per meter along the kiln axis, of:

$$\phi_{net,\,P} = 446732 + 361167 = 807899 \text{ W.m}^{-1}$$

The fraction of the thermal power transmitted via the regeneration effect is:

$$\frac{361167}{807899} = 44.7\%$$

This calculation may be easily programmed using a computer.

2

Tubular Heat Exchangers and Flat-plate Heat Exchangers

2.1. Description of exchangers

2.1.1. *Possible configurations*

There are three main types of shell-and-tube heat exchanger:

1) Fixed plate tubular exchangers:

The shell is welded to the two tube sheets, and the space between the bundle of tubes and the shell is minimized, improving heat transfer. However, the exterior of the tubes is not accessible. On the shell side, cleaning can only be carried out by chemical means.

When there is a significant difference between the temperature of the hot liquid and that of the cold liquid, the shell and tubes dilate and contract in different ways, creating inadmissible mechanical constraints. This problem may be overcome by the inclusion of expansion compensators in the shell cover, but this solution ceases to be possible at high pressure.

2) U-tube heat exchangers:

The use of tubes bent into a hairpin shape removes the need for one of the tube sheets, and allows the tube bundle to dilate and contract freely with changes in temperature. The bundle of tubes may be extracted from the shell for cleaning. However, the shape of the tubes makes internal cleaning difficult.

This configuration is suitable for use at high temperatures and with high pressures, and is reasonably cost-effective.

3) Floating head heat exchangers:

In this case, one of the tube sheets is fixed between the shell flange and the head flange. At the other end, the floating head is located within the shell and is able to move within the shell as the tubes dilate and contract. Both sides of the heat transfer surface can be cleaned by mechanical means.

This type of configuration is only possible for an even number of tube passes.

In a floating head configuration, there is a risk of a significant proportion of the fluid flow passing between the tube bundle and the shell as the space between the bundle and the shell is larger when compared with a fixed-head mechanism. In this case, longitudinal baffles are inserted along each side of the tube bundle to prevent the passage of fluid. One longitudinal baffle is generally used for five rows of tubes on each side of the bundle.

2.1.2. *Tube characteristics*

American standards are generally used in defining the thickness and diameter of exchanger tubes, and thicknesses are determined using the Birmingham Wire Gage (BWG).

The BWG ranges from 8 to 18 in the order of decreasing thickness. The table below shows equivalent thicknesses in mm.

8	9	10	11	12	13	14	15	16	17	18
4.19	3.76	3.40	3.05	2.77	2.415	2.101	1.83	1.65	1.475	1.245

The external diameter of the tubes can take one of the five possible values (expressed in inches):

1/2"   3/4"   1"   1" 1/4   1" 1/2

3/4" and 1" tubes are the most widely used.

Tubes may be laid out in a square or triangular pattern, as shown below:

Square layout Triangular layout

Figure 2.1. *Tube layouts*

The ratio between the pitch (distance from center to center) and the external diameter of the tubes is 1.25 or, more rarely, 1.33.

A square layout is preferred in cases where the space between the tubes needs to be cleaned using jets of steam.

Detailed geometric characteristics of exchanger tubes are given in Appendix 1.

Finally, note that the length of these tubes is generally a multiple of 50 cm.

The table below gives a number of layout values associated with tube diameters, all expressed in inches.

Square layout			Triangular layout		
d	p	p/d	d	p	p/d
3/4	1	1.33	3/4	15/16	1.25
1	1 ¼	1.25	3/4	1	1.33
1 ¼	1 9/16	1.25	1	1 ¼	1.25
1 ½	1 7/8	1.25	1 ¼	1 9/16	1.25

2.1.3. *Shell diameter and tube bundle diameter. Number of tubes*

A nominal space is left between the tube bundle and the shell in order to make the system easier to disassemble. The size of the space depends on the type of head being used.

Let D_c be the internal diameter of the shell, expressed in meters. Spacing becomes progressively larger as we move from a fixed head to a split ring system and from a split ring system to a pull-through system. More precisely:

Fixed head:

$D_c > 0.76$ m	$J_{TF} = 1.52$ cm
$D_c < 0.76$ m	$J_{TF} = 2\ D_c$ cm

Split ring (floating head):

$D_c < 0.25$ m	$J_{AF} = 2.5$ cm
$D_c > 1.3$ m	$J_{AF} = 5$ cm
$0.25\ m < D_c < 1.3$ m	$J_{AF} = 2.4\ D_c + 1.9$ cm

Pull-through (floating head):

$D_c < 0.25$ m	$J_{EC} = 7.5$ cm
$D_c > 1.3$ m	$J_{EC} = 11.5$ cm
$0.25\ m < D_c < 1.3$ m	$J_{EC} = 3.8\ D_c + 6.55$ cm

Given the diameter of the shell, an easy way to obtain the diameter of the tube bundle might be to subtract twice the nominal spacing.

In reality, however, if we wish to obtain a bundle diameter suitable for use in calculating a number of tubes, we must begin by calculating the diameter of a fixed-head shell corresponding to the same tube bundle.

For example, let D_c be the diameter of a split-ring shell, J_{AF} the corresponding nominal spacing and J_{TF} the nominal spacing of a fixed-head shell of the same diameter. The diameter of the equivalent fixed-head shell is:

$$D_{cTF} = D_c - 2(J_{AF} - J_{TF}) \qquad \text{(in coherent units)}$$

The effective diameter of the tube bundle is then obtained using the following formula:

$$D_F = D_{cTF} - 0.013 - p(1 + D_{cTF})$$

This formula is valid for both square and triangular layouts, on the condition that D_F, D_{cTF} and p (pitch: distance between centers of tubes) are expressed in meters.

It is now possible to evaluate the effective cross-section, from which we can determine the appropriate number of tubes:

$$A = \frac{\pi}{4} D_F^2 - 0.01 \times D_F \times N_{PT}$$

where N_{PT} is the number of tube passes. In this formula, each baffle is considered to take the place of the tubes over a width of 2 cm. The cross-section A is expressed in m^2 and D_F in m. In the case of a single pass, $N_{PT} = 0$.

Furthermore, the unitary cross-section occupied by a tube is:

$$-\text{for a square layout}: \; A_u = p^2$$

$$-\text{for a triangular layout}: \; A_u = p^2 \frac{\sqrt{3}}{2}$$

Finally, the number of tubes N_T is:

$$N_T = A/A_u$$

2.1.4. *Transversal baffles (characteristics)*

The spacing between tubes and baffles is generally 0.4 mm.

This spacing can be 0.2 mm for unsupported tubes over a length greater than 1 m.

The spacing between baffles and the shell is given by the table shown below, based on standards issued by the TEMA (Tubular Exchanger Manufacturers Association).

Internal diameter of shell (mm)	Spacing between baffle and shell (mm)
200–350	1.3
350–450	1.6
450–600	1.9
600–1,000	2.2
1,000–1,400	2.9
1,400 and above	3.8

This spacing results in a leakage flow, which modifies the values of the transfer coefficient and the pressure drop on the shell side.

The open side of a baffle is known as the window. Let X_{0C} represent the aperture of the baffle, i.e. the relationship between the deflection of the window and the diameter of the shell.

X_{0C} is generally equal to 0.25, but may be as high as 0.40 for a gas under reduced pressure, or as low as 0.15 if the flow on the baffle side is low.

2.2. Calculations on the tube side

2.2.1. *Heat transfer inside the tubes (clear liquid or gas)*

This method is also suitable for slurries obtained using particles with dimensions less than 60 μm.

The expression of the partial transfer coefficient differs depending on whether the flow regime is turbulent or laminar.

1) Turbulent (Re > 2,100):

The Nusselt number is obtained using:

$$\mathrm{Nu} = 0.027\ \mathrm{Re}^{0.8}\mathrm{Pr}^{1/3}\left(\mu / \mu_p\right)^{0.14}$$

μ: average viscosity of the fluid (Pa.s)

μ_p: viscosity of the fluid at wall temperature (Pa.s).

Re is the Reynolds number:

$$Re = \frac{V d_i \rho}{\mu}$$

V: velocity of the fluid ($m.s^{-1}$)

d_i: internal diameter of the tube (m)

ρ: density of the fluid ($kg.m^{-3}$).

Pr is the Prandtl number:

$$Pr = \frac{C_p \mu}{\lambda}$$

C_p: thermal capacity of the fluid ($J.kg^{-1}.°C^{-1}$)

λ: thermal conductivity of the fluid ($W.m^{-1}.°C^{-1}$).

The transfer coefficient for the inside of the tubes is then obtained using:

$$\alpha_i = \frac{Nu\lambda}{d_i}$$

If the fluid is water, then the coefficient 0.027 should be replaced by 0.023.

2) Laminar regime (Re < 2,100):

The Nusselt number is obtained using:

$$Nu = 1.49\left[Re\ Pr\frac{d_i}{L}\right]^{1/3}\left[1+0.015\ Gr^{1/3}\right]\left(\mu/\mu_p\right)^{0.14}$$

L: length of the tube (m).

Gr is the Grashof number:

$$Gr = \frac{d_i^3 \rho^2 g\, \beta |\Delta T|}{\mu^2}$$

g: acceleration due to gravity (9.81 m.s^{-2})

$|\Delta T|$: temperature difference between the wall and the fluid (°C)

β is the volumetric dilation coefficient. It is measured in °C^{-1} and is equal to 1/273 for a perfect gas. For liquids and gases at high pressure, it is simply deduced from the variation in bulk density as a function of the temperature. The grouping $\beta\ |\Delta T|$ can be clearly written:

$$\beta |\Delta T| = \left| \frac{v - v_p}{v} \right|$$

where v and v_p are the bulk densities of the fluid at its average temperature and at the temperature of the wall, respectively.

EXAMPLE 2.1.–

$V = 1.5$ m.s^{-1}	$\mu = \mu_p = 0.011$ Pa.s	$C_p = 3\ 344$ K.kg^{-1}.°C^{-1}
$d_i = 0.021$ m	$\rho = 1\ 100$ kg.m^{-3}	$\lambda = 0.14$ W.m^{-1}.°C^{-1}

$$Re = \frac{1.5 \times 0.021 \times 1100}{0.011} = 31200 > 2100$$

$$Pr = \frac{3344 \times 0.011}{0.14} = 263$$

$$Nu = 0.027 \times (31200)^{0.8} \times 263^{1/3} = 681$$

$$\alpha_i = \frac{681 \times 0.14}{0.021} = 4540 \text{ W.m}^{-2}.°C^{-1}$$

2.2.2. *Heat transfer inside the tubes (liquid-solid dispersion)*

The method described below is only suitable for particle sizes greater than 100 μm, and is based on the work carried out by Plass and Molerus [PLA 74].

The partial transfer coefficient on the dispersion side is calculated as if the sole liquid were considered, and then associated with a correction factor F_s to account for the presence of the solid. Hence:

$$\alpha_{LS} = F_s \alpha_L$$

Plass and Molerus' work [PLA 74] can be summarized analytically using the following expression (valid for turbulent regimes):

$$F_s = 1 + \frac{a\left(C_v C_{vo} - C_v^2\right)}{1 + bC_v}$$

C_v is the volume fraction of the solid in the dispersion and a, b and C_{vo} are parameters dependent on the Reynolds number of the flow and the diameter of the solid particles.

More precisely, we may calculate an optimal particle diameter d_{op}:

$$d_{op} = 127 D_T x^{0.625} Re^{-0.875} \left[\frac{\rho_L}{\rho_S}\right]^{0.625} \qquad (m)$$

D_T: internal diameter of tubing (m)

x: von Karman constant (0.4 for water)

ρ_L and ρ_S: densities of the liquid and the solid (kg.m^{-3}).

Re: Reynolds number

$$Re = \frac{VD_T\rho_L}{\mu}$$

μ: viscosity of the liquid (Pa.s)

V: speed of the suspension in the tubing ($m.s^{-1}$).

We then calculate:

$$X = d_p / d_{op}$$

d_p: diameter of particles (m)

$$\text{If } X < 1 \quad F_{s\,max} = 1 + 0.34\sqrt{X} \quad \text{and} \quad C_{vo} = 0.27$$

$$\text{If } X \geq 1 \quad F_{s\,max} = 1 + 0.34 X^{0,81} \quad \text{and} \quad C_{vo} = 0.17$$

In both cases, we take:

$$C_{v\,max} = 0.038$$

It is then possible to deduce parameters a and b:

$$b = \frac{1}{C_{v\,max}^2}\left(C_{vo} - C_{v\,max}\right)$$

$$a = \frac{\left(F_{s\,max} - 1\right)\left(1 + bC_{v\,max}\right)}{C_{vo}C_{v\,max} - C_{v\,max}^2}$$

The expression of F_s is therefore:

$$F_s = 1 + \frac{a\left(C_v C_{vo} - C_v^2\right)}{1 + b\,C_v}$$

The authors indicate that the optimum diameter d_{op} is simply 0.3 times the size of the eddies existing near the wall. Diameter d_{op} is that which results in the greatest increase in the transfer coefficient.

EXAMPLE 2.2.–

Consider the case of a loop exchanger in a forced circulation crystallizer. The crystal suspension, containing 15% by mass of crystals, circulates through the tubes.

$V = 2\ m.s^{-1}$ $D_T = 0.035\ m$ $\mu = 0.3 \times 10^{-3}\ Pa.s$

$\rho_L = 1{,}000\ kg.m^{-3}$ $\rho_s = 2{,}600\ kg.m^{-3}$ $d_p = 200 \times 10^{-6}\ m$

$$Re = \frac{2 \times 0.035 \times 1000}{0.3 \times 10^{-3}} = 2.3 \times 10^5$$

$$d_{op} = 127 \times 0.035 \times 0.4^{0.625} \times \left(2.3 \times 10^5\right)^{-0.875} \times \left(\frac{1}{2.6}\right)^{0.625}$$

$$d_{op} = 27.5 \times 10^{-6}\ m$$

$$X = 200/27.5 = 7.25 > 1$$

$$F_{s\,max} - 1 = 0.34/7.25^{0.81} = 0.068$$

$$C_{vo} = 0.17$$

$$b = \frac{1}{(0.038)^2}(0.17 - 2 \times 0.038) = 65.1$$

$$a = \frac{0.068(1 + 0.038 \times 65.1)}{0.17 \times 0.038 - 0.038^2} = 47.05$$

$$F_s = 1 + \frac{47.05\left(0.17C_v - C_v^{\ 2}\right)}{1 + 65.1C_v}$$

$$C_v = \frac{0.15/2600}{\dfrac{0.85}{1000} + \dfrac{0.15}{2600}} = 0.0635$$

$$F_s = 1 + \frac{47.05\left(0.17 \times 0.0635 - 0.0635^2\right)}{1 + 65.1 \times 0.0635}$$

$$F_s = 1.062$$

The presence of crystals can be seen to slightly improve the coefficient α_{LS}. However, *if C_v is greater than 0.17, the coefficient α_{LS} drops off extremely quickly.* For this reason, the acceptable volume fraction for crystals in crystallization systems is limited, as these systems almost always make use of heat transfer mechanisms.

2.2.3. *Pressure loss inside the tubes*

This pressure loss is obtained using:

$$\Delta P = f\frac{L}{d_i} \times \frac{\rho\, V^2}{2}$$

The notation is the same as that used previously. ΔP is given in Pa, and f is calculated using:

– Laminar regime: (Re < 2,100)

$$f = 72/Re$$

– Turbulent regime: (Re ≥ 2,100)

$$f = 0.007 + \frac{0.625}{Re^{0.32}}$$

Input, output and turning from one pass to the next should also be taken into consideration. For each of these three operations, the pressure loss is:

$$\Delta P = \frac{4\rho V^2}{2}$$

Finally, the pressure loss in the tubes of an exchanger with n passes is:

$$\Delta P = \frac{n\rho V^2}{2}\left(4 + f\frac{L}{d_i}\right) \qquad \text{(where n is the number of passes)}$$

EXAMPLE 2.3.–

$L = 3$ m $\qquad V = 1.5$ m.s^{-1} $\qquad \mu = 10^{-3}$ Pa.s

$d_i = 0.021$ m $\qquad \rho = 1{,}000$ kg.m^{-3} $\qquad n = 4$

$$Re = \frac{1000 \times 1.5 \times 0.021}{10^{-3}} = 31500 > 2100$$

$$f = 0.007 + \frac{0.625}{(31500)^{0.32}} = 0.0297$$

$$\Delta P = \frac{4 \times 1000 \times 1.5^2}{2}\left(4 + 0.0297 \times \frac{3}{0.021}\right) = 37106 \text{ Pa}$$

Giving a liquid height of:

$$\frac{\Delta P}{\rho g} = \frac{37106}{1000 \times 9.81} = 3.78 \text{ m C.L.}$$

2.3. Calculations on shell side

2.3.1. *Gross transfer coefficient, shell side (Bell method [BEL 60])*

This transfer coefficient includes corrections for:

– the current deflected around the tube bundle;

– the openness of the buffers;

– the number of rows of tubes traversed by a crossflow current.

The overall cross-section traversed by a crossflow current:

$$A_c = L_C\left[D_c - D_F + \frac{(D_F - d)}{P_N}(P_N - d)\right]$$

A_c: m^2

L_C: shell diameter (m)

D_F: effective diameter of the tube bundle (m)

d: external diameter of the tubes (m)

P_N: pitch of tubes affected by the crossflow current (m)

(P_N is identical to p in square and classic triangular layouts).

The flowrate density for section A_c is (Bell gives a slightly different method in his publication from 1960):

$$G = \frac{W}{A_c}$$

G: flowrate density ($kg.m^{-2}.s^{-1}$)

W: flowrate ($kg.s^{-1}$).

The Reynolds number for the crossflow current is:

$$Re = \frac{Gd}{\mu}$$

This is seen in the element j in Bell's Figures 5 and 2 [BEL 60].

μ: viscosity of liquid (Pa.s).

It is now necessary to calculate the correction required to account for *the current deflected around the tube bundle.* The fraction of A_c attributed to deflected current is:

$$F_D = \frac{(D_C - D_F)L_C}{A_c}$$

The deflected current encounters longitudinal baffles, which send it back into the tube bundle. We shall consider that one baffle is placed on each side of the tube bundle for every five rows of tubes.

The number of longitudinal baffles per row of tubes N_B is thus equal to 0.2.

More generally, *the correction factor for the deflected current* is, according to Bell (equation [2] in [BEL 60]):

$$C_{CD} = e^{-m}$$

$$m = a\, F_F \left(1 - \sqrt[3]{2N_B}\right) \quad \text{if } N_B \leq 0.5$$

$$m = 0 \quad \text{if } N_B > 0.5$$

In this expression:

$$a = 1.25$$

We shall now calculate the correction needed to account for the opening of the baffles:

Let X_{oc} be the buffer aperture, i.e. the ratio between the baffle deflection and the diameter of the shell:

X_{oc} varies from 0.15 to 0.40.

The deflection of the baffle window is thus $\ell_c = D_c\, X_{oc}$

We now need to *evaluate the fraction F_F of the tube bundle located within a window.* This section is the difference between the sector

$$\frac{1}{2} R_F^2 \times 2\,\theta = \theta R_F^2$$

and two triangles:

$$2 \times \frac{HR_F}{2} = HR_F$$

where R_F is the effective radius of the tube bundle, i.e. $D_F/2$.

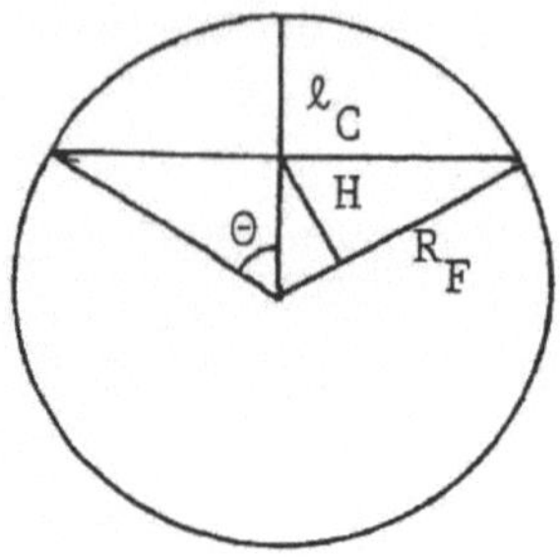

Figure 2.2. *Baffle window*

The cross-section of the tube bundle is πR_F^2, hence

$$F_F = \frac{\theta}{\pi} - \frac{H}{\pi R_F}$$

with:

$$\theta = \arccos\frac{(D_c/2 - l_c)}{D_F/2} = \arccos\frac{(D_c/2 - 2l_c)}{D_F}$$

$$H = (D_c/2 - l_c)\sin\theta$$

Thus:

$$F_F = \frac{1}{\pi}\left[\arccos\frac{(D_c - 2l_c)}{D_F} - \frac{(D_c - 2l_c)}{D_F}\sin\left[\arccos\frac{(D_c - 2l_c)}{D_F}\right]\right]$$

The correction coefficient for the aperture of the baffles C_{oc} is given by Bell's equation [12b] (in [BEL 60]):

$$C_{oc} = 1 - F_F + 0.524 F_F^{0.32}$$

Let us now evaluate the *correction for the number of rows of tubes with a crossflow current.*

This number is:

$$N_{RT} = \frac{D_c\left(1-2X_{oc}\right)}{p_p}$$

p_p: pitch of tubes counted parallel to the fluid crossflow current

$\left(\text{for triangular spacing, } p_p = p\sqrt{3/2}\right)$.

The coefficient varies from 0.66 to 1.05 as N_{RT} varies from 1 to 20 and above. This will not be taken into consideration.

The final gross transfer coefficient is therefore:

$$\alpha_b = \frac{jC_pG\,C_{CD}C_{oc}}{\left(\mu_p/\mu\right)^{0.14}\,Pr^{2/3}}$$

j: coefficient given by Figure 5 [BEL 60]

μ_p: viscosity at the temperature of the wall (kg.sm)

Pr: Prandtl number of the fluid.

2.3.2. *Coefficient correction for leakage*

Leakage occurs between the tubes and the baffles and between the baffles and the shell.

If j_{TC} is the spacing between the tubes and the baffles, the area of leakage is:

$$A_{TC} = j_{TC}\pi dN_T\left(1-F_F\right)$$

In addition, with J_{CC}, the spacing between the baffle and the shell, the corresponding leakage area is:

$$A_{CC} = J_{CC} D_c \left[\pi - \text{Arc}\cos\left(1 - \frac{2l_c}{D_c}\right)\right]$$

The longitudinal leakage cross-section is:

$$A_L = A_{TC} + A_{CC}$$

Our corrected coefficient is then obtained using the formula:

$$\alpha_c = \alpha_b \left[1 - \frac{A_{TC} + 2A_{CC}}{A_L} \left\{ 0.45 \frac{A_L}{A_c} + 0.1 \left[1 - \exp\left(-30 \frac{A_L}{A_c} \right)\right]\right\}\right]$$

Readers may wish to consult Perry, or Bell [BEL 60], to obtain the correction applied to the transfer coefficient.

2.3.3. *Pressure loss due to crossflow current*

This pressure loss is obtained using the following expression:

$$\Delta P_c = \frac{4fN_{RT}G^2}{2\rho} C'_{CD} \left(\mu_p/\mu\right)^{0.14}$$

ℓ: fluid density (kg.m^{-3})

$$C'_{CD} = e^{-m} \quad \text{with} \quad m = aF_D\left(1 - \sqrt[3]{2N_B}\right)$$

$$\text{If Re} < 100 \quad a = 5$$

$$\text{If Re} \geq 100 \quad a = 4$$

f: friction coefficient.

Pressure loss through baffle windows (Bell's method):

To evaluate this pressure loss, we need to know the area of the cross-section, which is left empty between the tubes for the window.

By analogy with the calculation of F_F, the surface of a window, without tubes, can be seen to be:

$$A_{FV} = \frac{D^2c}{4}\left[\text{Arc}\cos\left(1-\frac{2l_c}{D_c}\right)-\left(1-\frac{2l_c}{D_c}\right)\sin\left[\text{Arc}\cos\left(1-\frac{2l_c}{D_c}\right)\right]\right]$$

The cross-section occupied by tubes passing through the window is:

$$A_{FT} = N_T\frac{\pi d^2}{4}\times F_F$$

The area of free passage through the window is:

$$A_F = A_{FV} - A_{FT}$$

Hence, the bulk flow density through the window is:

$$G_F = \frac{W}{A_F}\text{kg.m}^{-2}.\text{s}^{-1}$$

The geometric average of the crossflow and parallel currents is:

$$V_m = \frac{\sqrt{GG_F}}{\rho}\left(\text{m.s}^{-1}\right)$$

The number of rows of tubes in the window of a baffle is obtained using the following semi-empirical expression:

$$N_{RF} = \frac{0.8}{p_p}l_c$$

It is now possible to calculate the pressure loss in a window.

Depending on the value of the Reynolds number:

$$\mathrm{Re} = \frac{G_F d}{\mu}$$

if $\mathrm{Re} \leq 100$

$$\Delta P_F = \frac{2\rho V_m^2}{2} + V_{m\mu}\mu\left(\frac{28 N_{RF}}{p-d} + \frac{26}{d_{eq}}\frac{L_c}{d_{eq}}\right)$$

with an equivalent diameter:

$$d_{eq} = \frac{4A_F}{P_F}$$

P_F is the wetted perimeter of the shell and the tubes in the window.

$$P_F = \frac{D_c}{2} \times 2 \text{ arc cos}\left(1 - \frac{2l_c}{D_c}\right) + N_T \pi d F_F$$

If $\mathrm{Re} > 100$

$$\Delta P_F = \left(2 + 0{,}6 N_{RF}\right)\frac{\rho V^2 m}{2}$$

2.3.4. *Pressure drop: shell side*

From the discussion above, we deduce:

$$\Delta P_{TOT} = kN_c\left(\Delta P_c + \Delta P_F\right) + 2\Delta P_c\left(1 + \frac{N_{RF}}{N_{RT}}\right)$$

In this expression:

N_c is the number of transversal baffles:

$$N_c = \frac{L_T}{L_c} - 1$$

k is a correction coefficient that takes account of losses between tubes and baffles and between baffles and the shell:

$$k = 1 - \frac{A_{TC} + 2A_{CC}}{A_L} \times \left[0.57 \frac{A_L}{A_c} + 0.27 \left[1 - \exp\left(-20 \frac{A_L}{A_c} \right) \right] \right]$$

Bell [BEL 60] or Perry may also be consulted in order to calculate the pressure drop.

Numerical example of calculation from the shell side:

Consider the following example:

100,000 $kg.h^{-1}$ of methyl alcohol needs to be cooled from 93° to 38° using water heated from 27° to 38°. The tubes used are 15.4/19.05, i.e. 3/4'' BWG 15. These tubes are laid out in a triangular pattern with a spacing of 24 mm and have a length of 5 m.

The shell is a split-ring, floating head model with an internal diameter of 820 mm.

The shell is equipped with transversal baffles, opened at 16% of the diameter, with a spacing of 160 mm.

Solution:

$$j_{AF} = 2.4 \times 0.82 + 1.9 = 3.9 \text{ cm} = 0.039 \text{ m}; \; j_{TF} = 1.52 = 0.0152 \text{ m}$$

$$D_{cT} = 0.82 - 2(0.039 - 0.0152) = 0.773$$

$$D_F = 0.773 - 0.013 - 0.024(1 + 0.77) = 0.7175 \text{ m} \# 0.72 \text{ m}$$

$$N_T = \frac{\pi}{4} \times 0.72^2 / \left(0.024^2 \times \sqrt{3} / 2 \right) = 816 \text{ tubes}$$

$$A_c = 0.16 \left[0.82 - 0.72 + \frac{(0.72 - 0.019)}{0.024} (0.024 - 0.019) \right]$$

$$A_c = 0.04\text{m}^2$$

$$G = \frac{100000}{0.04 \times 3600} = 694 \text{ kg.m}^{-2}.\text{s}^{-1}$$

$$\text{Re} = \frac{694 \times 0.019}{0.32 \times 10^{-3}} = 41300 \quad j = 0.005 \qquad \text{(see Appendix 1)}$$

$$F_D = \frac{(0.82 - 0.72)0.16}{0.04} = 0.40$$

$$m = 1.35 \times 0.40 \times \left(1 - \sqrt[3]{2} \times 0.2\right) = 0.14$$

$$C_{CD} = 0.87$$

$$l_c = 0.16 \times 0.82 = 0.13 \text{ m}$$

$$F_F = \frac{1}{\pi}\left[\text{Arc}\cos\left(\frac{0.82 - 0.13 \times 2}{0.72}\right) - \left(\frac{0.82 - 2 \times 0.13}{0.72}\right)\sin\left[\text{Arc}\cos\left(\frac{0.82 - 2 \times 0.13}{0.72}\right)\right]\right]$$

$$\text{Arc}\cos\left(\frac{0.82 - 2 \times 0.13}{0.72}\right) = 0.68 \text{ rad}$$

$$F_F = \frac{1}{3.14}\left[0.68 - 0.78\sin(0.68)\right] = 0.06$$

$$C_{oc} = 1.15$$

$$N_{RT} = \frac{0.82(1 - 2 \times 0.16)}{0.024 \times \frac{\sqrt{3}}{2}} = 65$$

$$C_{RT} = 1.06$$

$$\alpha_b = \frac{0.005 \times 2842 \times 694 \times 0.87 \times 1.15 \times 1.06}{1.02 \times 2.83}$$

$$\alpha_b = 3619 \text{ W.m}^{-2}.°\text{C}^{-1}$$

$$A_{TC} = 4\times10^{-4}\times\pi\times0.019\times816\times0.94$$

$$A_{TC} = 0.0183 \text{ m}^2$$

$$A_{CC} = 2.2\times10^{-3}\times0.82\left[\pi - \text{arc cos}(1-0.32)\right] = 0.0042 \text{ m}^2$$

$$A_L = 0.0183 + 0.0042 = 0.0225 \text{ m}^2$$

$$\alpha_c = 3619\left[1-\frac{0.0183+2\times0.0042}{0,0225}\left\{0.45\times\frac{0.0225}{0.04}+0.1\left[1-\exp-\left(30\times\frac{0.0225}{0.04}\right)\right]\right\}\right]$$

$$\alpha_c = 2531 \text{ W.m}^{-2}.°\text{C}^{-1}$$

$$C'_{CD} = \exp-\left[4\times0.4\left(1-\sqrt[3]{2\times0.2}\right)\right]$$

$$C'_{CD} = 0.66$$

$$\Delta P_c = \frac{4\times0.1\times65\times694^2\times0.66\times1.02}{2\times775}$$

$$\Delta P_c = 5438 \text{ Pa}$$

$$A_{FV} = \frac{0.82^2}{4}\left[\text{Arc cos}(1-2\times0.16)-(1-2\times0.16)\sin\left[\text{Arc cos}(1-2\times0.16)\right]\right]$$

$$A_{FV} = \frac{0.82^2}{4}(0.82-0.68\times0.73) = 0.0544$$

$$A_{FT} = \frac{816\times3.14\times0.019^2\times0.06}{4} = 0.0139$$

$$A_F = 0.0544 - 0.0139 = 0.0405$$

$$G_F = \frac{100000}{0.0405 \times 3600} = 686 \text{ kg.m}^{-2}.\text{s}^{-1}$$

$$V_m = \frac{\sqrt{686 \times 694}}{775} = 0.89 \text{ m.s}^{-1}$$

$$N_{RF} = 0.8 \times \frac{0.131}{0.021} = 5$$

$$\Delta P_F = (2 + 0.6 \times 5)\frac{775 \times 0.89^2}{2} = 1535 \text{ Pa}$$

$$N_c = \frac{5}{0.16} - 1 = 30$$

$$k = 1 - \frac{0.0183 + 2 \times 0.0042}{0.0225}\left[0.57 \times \frac{0.0225}{0.04} + 0.27\right]$$

$$k = 0.30$$

$$\Delta P_{TOT} = 0.30 \times 30(5438 + 1535) + 2 \times 5438\left(1 + \frac{5}{65}\right)$$

$$\Delta P_{TOT} = 74500 \text{ Pa}$$

$$\Delta P_{TOT} = 7.45 \text{ m CE}$$

2.3.5. *Overall heat transfer coefficient*

This coefficient, which we shall call U, is given by the relationship:

$$\frac{1}{U} = \frac{1}{\alpha_i}\frac{d_e}{d_i} + R_{fi}\frac{d_e}{d_i} + \frac{e}{\lambda} + \frac{1}{\alpha_e} + R_{fe}$$

α_i: transfer coefficient for the inside of the tubes (W.m^{-2}.°C^{-1})

R_{fi}: fouling resistance inside the tubes (m^2.°C.W^{-1})

d_e: external diameter of tubes (m)

d_i: internal diameter of tubes (m)

e: wall thickness (m)

λ: wall conductivity ($W.m^{-1}.°C^{-1}$)

α_e: transfer coefficient for the outside of the tubes ($W.m^{-2}.°C^{-1}$)

R_{fe}: fouling resistance on the outside of the tubes ($m^2.\ °C.W^{-1}$).

Note that the coefficient of an exchanger is always calculated in relation to *the external surface of the tubes.*

For the transmitted heat flow in a tube to remain constant, coefficient α must be inversely proportional to the cross-section of the flow tube; this explains the presence of a correction factor $\frac{d_e}{d_i}$ applied to $1/\alpha_i$ and R_{fi}.

By eliminating the fouling resistances R_{fi} and R_{fe}, we obtain the overall "clean" coefficient U_p.

2.4. Practical data

2.4.1. *Usual pressure drop values*

The acceptable level of power loss for a tubular exchanger is lower than that for a compact exchanger.

Generally, a value of 3–5 m CE on the shell side and 5–7 m CE on the tube side is considered acceptable, and the baffles are opened to 25% of the diameter.

When the pressure of the fluid is lower than the atmospheric pressure, a loss between 1/5 and 1/10 of the pressure is considered normal. In this case, the baffles are opened to 40% of the diameter.

2.4.2. *Computerized exchanger calculations*

The methods discussed above can be used to simulate the performance of an exchanger where the nature, flow rate and temperature of the exchanger are known.

However, identification of a suitable device for specific performance expectations requires an iterative calculation, which may be carried out as follows:

First, the nature of the tubes, i.e. diameter and thickness, must be selected.

Then, the required length of the tubes is selected. This length is an increasing function of the length of the heat exchange, the available space and, evidently, the thermal power. This final point simply means that there must be a reasonable aspect ratio between the diameter of the shell and the length of the tubes; for example, a shell with a diameter of 1.5 m should not be used with tubes of 2 m length.

The computer program will calculate an appropriate system, i.e. identify the number of tubes required for a number of passes on the tube side increasing from 1 to 8:

$$1-2-4-6-8.$$

The number of passes for which the surface is smallest will be selected.

The remaining task involves calculating a system for tubes of given dimensions and for a given number of passes. To do this, we calculate the number of tubes corresponding to the imposed maximum pressure loss on the tube side. Using this number of tubes, we then determine the number of baffles required to satisfy the maximum acceptable pressure loss on the shell side. We then calculate the overall coefficient U and the exchange surface S. Taking account of the MLTD (mean logarithmic temperature difference) for the desired temperature program, the calculated heating capacity is:

$$Q_c = US\Delta T$$

Two situations are possible, depending on whether this capacity is above or below the heating power to transfer Q_N, imposed by the heating program:

1) $US\Delta T > Q_N$

The transfer surface needs to be reduced, but this would result in an increase in the pressure drop on the tube side, which is not possible.

The number of passes under consideration corresponds to an over-dimensioned device.

2) $US\Delta T < Q_N$

In this case, the number of tubes may be increased until an adequate surface is obtained, but for this number of passes, the pressure drop is not fully exploited.

Now, consider the evolution of the previous conclusions with an increasing number of passes, from 1 to 8.

If the number of passes is low, few tubes are needed in order to reach the desired pressure drop limit on the tube side.

This falls within the scope of situation 2, and the pressure drop must simply be reduced in order to increase the number of tubes, i.e. $US\Delta T$ (it can, in fact, be shown that, for any flow regime, product US is a decreasing function of the pressure drop as the number of tubes varies).

As the number of passes increases, the number of tubes required to respect the pressure drop limit also increases, and, for a given number of passes, the situation changes from 2 to 1. The point at which this change occurs is the optimum.

2.4.3. *Note: caloric temperature*

The notion of caloric temperature developed by Kern [KER 50, pp. 94–97] is only relevant for certain types of manual calculation (heavy hydrocarbons).

For computerized calculations, note simply that if there is a linear relationship between the overall coefficient U and the temperature of one of the two fluids, the transfer relationship is written:

$$\frac{Q}{S} = \frac{U_c \Delta T_f - U_f \Delta T_c}{\mathrm{Ln}\dfrac{U_c \Delta T_f}{U_f \Delta T_c}}$$

Indices c and f indicate that the hot and cold extremities of the exchanger are being considered when evaluating the overall coefficient U and the temperature difference ΔT between the hot and cold fluids.

When designing a system, this transfer relationship may be directly substituted for the classic relationship without creating any additional problems.

In the case of performance calculations, where the extreme temperatures are unknown, the use of this relationship requires additional iterations; this issue is not resolved by using an average temperature to evaluate U in the classic formula.

2.5. Plate exchangers

2.5.1. *Description*

A plate heat exchanger is made up of a pack of corrugated plates, held between two plates of stainless steel using a suitable holding system. The plates are designed in such a way as to allow easy cleaning and to avoid catching. The liquid circulation areas are separated by a single molded joint, placed inside a specially designed groove.

The simplest form of plate grouping uses one pass for each side, and this configuration is used wherever possible. In this case, pressure loss at the beginning and end of each pass is eliminated. Furthermore, all of the piping is attached to the fixed frame, and does not need to be disassembled when the exchanger is opened for examination or cleaning purposes. In the case of condensation processes, there will be one pass on the vapor side, and multiple passes on the liquid side. In this situation, intermediate plates, equipped with input and output piping, are required for each pass.

Plate exchangers are easy to open (NB: this is not the same as disassembly), which is useful if:

– cleaning is required;

– the heating program has changed, and plates need to be added or removed.

2.5.2. *Number of transfer units*

Let T be the temperature of the hot liquid and t the temperature of the cold liquid. The input and output of each liquid are characterized by indices 1 and 2.

Let Δt_m be the MLTD:

$$\Delta t_m = \frac{(T_1 - t_2) - (T_2 - t_1)}{\text{Ln}\dfrac{T_1 - t_2}{T_2 - t_1}}$$

The number of transfer units (NTU) for each of the two fluids is:

$$\theta = \frac{t_2 - t_1}{\Delta t_m} \quad \text{and} \quad \Theta = \frac{T_1 - T_2}{\Delta t_m}$$

An exchange is said to be short if NTU < 1 and long if NTU > 1. In our discussion below, we shall only consider the case of a cold liquid; the situation would be identical for a hot liquid.

The exchanged thermal power is:

$$W = US\Delta t_m = Q\rho C(t_2 - t_1)$$

Q: flow volume ($m^3.s^{-1}$)

ρ: bulk density of the liquid ($kg.m^{-3}$)

C: bulk thermal capacity of the liquid ($J.kg^{-1}.°C^{-1}$)

S: heat transfer surface (m^2)

U: overall heat transfer coefficient ($W.°C^{-1}.m^{-2}$).

Hence:

$$\theta = \frac{t_2 - t_1}{\Delta t_m} = \frac{US}{Q\rho C}$$

Classic tubular exchangers are convenient where $\theta < 1$, i.e. for short exchanges. Plate exchangers are recommended for long exchanges ($\theta > 1$).

This is due to the fact that the overall coefficient U of plate exchangers is very high, and can reach values between 3,000 and 4,000 $W.m^{-2}.°C^{-1}$, almost 10 times more than that of a tubular exchanger.

2.5.3. *The pressure loss coefficient*

For a cold liquid, take:

$$J = \frac{\Delta P}{\theta} = \frac{Q\Delta P(\rho C)}{US}$$

ΔP: pressure drop for the cold liquid traveling through the exchanger.

Generally, the pressure drop parameter J *increases* as the coefficient U increases.

If the value of J is high, the plate is said to be "hard". If J is low, then the plate is said to be "soft".

Figure 2.3 shows that parameter J is higher for tubular exchangers than for plate exchangers with the same transfer coefficient. In other words, for a given temperature program and a given pressure drop, the transfer coefficient U is much higher for the plate exchanger.

On average, and for a given heating program (flows and temperatures), the pressure drop through a plate exchanger is 6–7 times lower than that of a tubular exchanger.

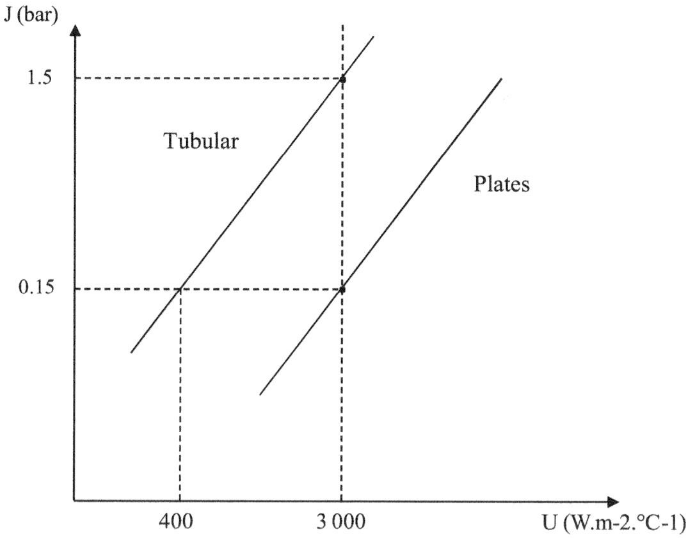

Figure 2.3. *Parameter J as a function of the transfer coefficient*

2.5.4. *Plate grouping and correction coefficient*

Using a plate exchanger, one or more passes may be carried out over each liquid. The MLTD correction coefficient F is, for example, with groupings 2/1, 3/1 and 4/1:

$$F \# 0.6 \text{ when } NTU = 8 \quad \text{and} \quad F \# 0.8 \text{ when } NTU = 3.5$$

Symmetrical groups: 1/1, 2/2, 3/3:

$$F \# 0.9 \text{ when } NTU = 4$$

$$F \# 0.8 \text{ when } NTU = 6$$

The manufacturer uses an automatic calculation program to determine the number and type of plates, along with their groupings for a given heating program.

2.5.5. *Fouling*

Fouling coefficients are given in Table 2.1.

Nature of fluid	Coefficient $10^4.m^2.°C.W^{-1}$
Demineralized water	0.1
Soft tap water	0.2
Hard tap water	0.5
Seawater	0.3–0.5
Oils (mineral or vegetable)	0.2–0.6
Organic products	0.1–0.3

(note that the thermal conductivity of lime-scale is around 0.5 $W.m^{-1}.°C^{-1}$).

Table 2.1. *Fouling coefficients*

The low level of fouling is due to:

– turbulence;

– the smooth surface of the plates;

– absence of dead space;

– use of noble materials for plates (no corrosion).

Plate exchangers are easy to open, giving access to the plates for cleaning purposes.

2.5.6. *Construction materials*

Except for the piping, no welding is required, with connections created by pressing; most of the noble materials are able to withstand this type of treatment (with the exception of pure zirconium, which is too brittle):

AISI 301, AISI 316, titanium, nickel, Monel, Incolloy, Hastelloy, copper, aluminum, silver, tantalum, etc.

Mild steel is not used in plate exchangers.

Widely available elastomers, such as natural rubber, ethylene, propylene, silicones, etc., are used for the joints. Teflon is unsuitable due to its visco-elastic behavior.

Ebonized piping may also be used, and is cheaper than noble alloys.

3

Finned Tube Exchangers

3.1. General points

3.1.1. *Purpose of finned tubes*

The heat transfer coefficient obtained by forced convection on a wall is considerably higher for a liquid than for a gas. This imbalance can be corrected by changing the form of the wall separating liquid and gas, so that the face in contact with the gas has a much larger surface area than the face in contact with the liquid. The most economical way of achieving this is through using tubes with transversal fins, i.e. fins with faces running perpendicular to the axis of the tube. The outline of the fins is generally circular.

3.1.2. *Types of finned tube exchangers*

Equipment used in kilns and boiler systems will not be discussed here. Within the chemical industry, there are three main types of finned tube devices:

– air coolers, in which atmospheric air is used to cool a liquid;

– air condensers, in which atmospheric air is used to condensate a mixed or pure vapor;

– steam heaters are a specific type of condenser; whilst the air in a classic condenser should not be heated above 60°C, steam heaters can supply a dryer with air at 170°C.

3.1.3. *Definition of the frontal area*

The tube bundle in a bare tube exchanger is cylindrical; when finned tubes are used, however, the bundle takes the form of a parallelepiped. Tubes are arranged in rows (or layers) perpendicular to the gas flow. The rectangular surface of each layer is of length L_T (length of the tubes) and width ℓ_R (the space between the two sides of the shell). The frontal area open for contact with the gas, i.e. the surface of a row is:

$$A_F = L_T \, \ell_R$$

ℓ_R: width of a row (m).

3.2. Technical data

3.2.1. *Bundle layout*

– Air coolers: the layers of tubes are placed on a 1–2% incline in order to allow draining to occur using gravity.

– Air condensers: tubes are placed on a horizontal incline of 5–10% to allow the condensate to flow. Devices installed in distillation columns use horizontal piping, which distributes vapor through finned tubes placed perpendicular to the horizontal pipe, with an incline on a horizontal level.

– Steam air duct heaters: the tube layers are installed vertically, and the tubes within these layers are also vertical to allow the condensate to flow.

3.2.2. *Air cooler header boxes*

Two main types of header boxes are used and they are as follows:

– with a cover;

– with plugs.

Covered header boxes are most widely used in the chemical industry, as it is possible to monitor and clean the inside of the box and the tubes without disassembling the tubes.

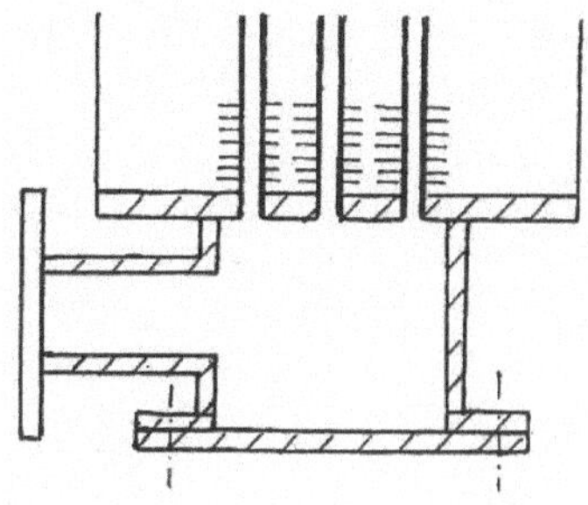

Figure 3.1. *Header box with a cover*

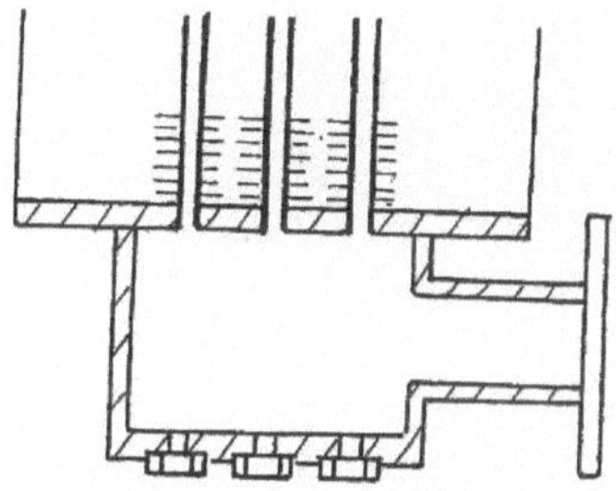

Figure 3.2. *Header box with plugs*

Plugged header boxes are used for higher pressures, as they are more airtight. They present all of the advantages of the covered header box, on condition that the plug holes are sufficiently large.

3.2.3. *Finned tube characteristics for air coolers*

In practice, tubes are always laid out in a triangular pattern.

Ext. diam. bare tube	Surface bare tube	Fin height	Fin thickness	Tube spacing	Fin diameter	Ratio
d_e (mm)	S_T (m^2/m)	h (mm)	e (mm)	p (mm)	d_a (mm)	p/d_a
25.4	0.08	12.7	0.4	54	51	1.06
25.4	0.08	15.9	0.4	67	57	1.18
38.1	0.12	15.9	0.4	83	70	1.18

Table 3.1. *Tube characteristics*

The fin density is characterized by a number N_a, representing the number of fins per meter, and by the spacing between two consecutive fins s (mm).

N_a	236	276	315	354	393	433
s	3.8	3.2	2.8	2.4	2.1	1.9

Table 3.2. *Fin spacing*

Fin density increases as the coefficient inside the tubes increases, so that the resistance on the air and process sides is of the same order of magnitude when observed from the exterior of the bare tubes.

The fin surface (both sides) per meter of the tube is:

$$S_a = 2N_a \frac{\pi}{4}\left(d_a^2 - d_e^2\right)$$

where, d_a and d_e are the diameter of the fins and the external diameter of the tubes.

The type of the tube is selected by identifying a fin density that increases as the overall coefficient U increases, hence the following table.

$U\left(W.m^{-2}.°C^{-1}\right)$	N_a
800	433
130	236

Table 3.3. *Fin density*

Furthermore, the diameter of the tubes is an increasing function of the size of the device. Finally, the thickness of the tubes (i.e. the BWG) is determined with regard to the pressure on the process side. A BWG of 14 is often sufficient.

Fins are often made from an aluminum sheathed tube. Aluminum can be used at temperatures of up to 300°. A 99.5% alloy is generally used as it is considered to offer better resistance to atmospheric corrosion.

Note that, in the case of steam air duct heating systems, mild steel fins are used.

The thermal conductivity of metals used in fin construction is given below.

Metal	$\lambda\left(W.m^{-1}.°C^{-1}\right)$	
	20°C	400°C
Mild steel	55	46
Aluminum	200	250
Copper	390	360

Table 3.4. *Conductance of materials*

3.3. Fin efficiency

3.3.1. *Preliminaries*

Our discussion will be limited to the case of circular transversal fins, which are most commonly used in practice. These fins have a constant thickness equal to 2e.

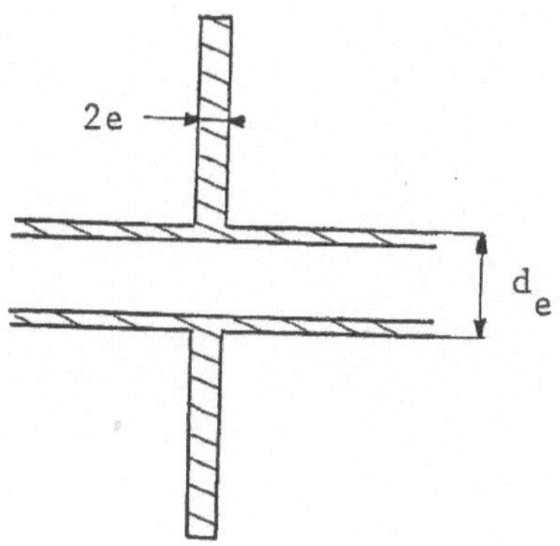

Figure 3.3. *Circular fin*

The base of the fins is at the same temperature as the exterior of the tube, t_e. The edges of the fins, however, are at a temperature t_a which is closer to the invariable temperature t_g of the gas. The transfer potential is therefore reduced, with a corresponding effect on the heat transfer.

3.3.2. *Temperature range in a fin*

Let α be the transfer coefficient between the surrounding gas and the metallic surface. This coefficient will be presumed to be uniform over the fins and the remaining bare surface of the tube.

As the fins have a circular crown shape, the surface element of a fin, for both faces and with an increment dr of distance r from the tube axis is:

$$dS_a = 4\pi r dr$$

Let t_m be the temperature (variable depending on position) of the fin metal, and take, in the case of heating tubes:

$$\theta = t_m - t_g$$

Let q be the heat flow propagating from the tube to the edge of the fins per meter of tube. This flow is *reduced* as we move away from the tube axis:

$$dq = -\alpha\theta dS_a = -4\pi\alpha\theta r dr$$

i.e.:

$$\frac{dq}{dr} = -4\pi\alpha\theta r \qquad [3.1]$$

The flow density φ may be expressed using Fourier's law:

$$\varphi = \frac{q}{2\pi r \times 2e} = -\lambda\frac{dt_m}{dr} = -\lambda\frac{d\theta}{dr}$$

or:

$$q = -4\pi re\lambda\frac{d\theta}{dr}$$

Deriving in relation to r:

$$\frac{dq}{dr} = -4\pi re\lambda \frac{d^2\theta}{dr^2} - 4\pi e\lambda \frac{d\theta}{dr} \tag{3.2}$$

By eliminating dq/dr in [3.1] and [3.2], we obtain:

$$\frac{d^2\theta}{dr^2} + \frac{1}{r}\frac{d\theta}{dr} - \frac{\alpha\theta}{\lambda e} = 0$$

Take:

$$x = r\sqrt{\alpha/\lambda e}$$

The previous equation becomes:

$$\frac{d^2\theta}{dx^2} + \frac{1}{x}\frac{d\theta}{dx} - \theta = 0$$

This is a *modified* Bessel equation (there is a negative sign θ) of order zero, for which the boundary conditions are:

1) At the base of the fin, the temperature of the metal is equal to the external temperature of the tube:

for $r = r_e$, $t_m = t_e$, i.e.: $\theta = \theta_e$

2) The heat transfer by conduction at the edge of the fin is considered to be zero. This hypothesis simplifies the analytical expression of the solution, and the associated error is minimum:

for $= r_a$, $d\theta/dx = 0$

The general solution to a zero-order modified Bessel equation is a linear combination of the modified zero-order Bessel functions of the first and second kind:

$$\theta = AI_0(x) + \beta K_0(x)$$

To satisfy the first condition, we simply write:

$$\theta = \theta_e \left[\frac{I_0(x) + \beta K_0(x)}{I_0(x_e) + \beta K_0(x_e)} \right] = C\left[I_0(x) + \beta K_0(x) \right]$$

It has been shown that (see Goudet [GOU 65]):

– $dI_0/dx = I_1$ (modified first-order Bessel function of the first kind)

– $dK_0/dx = K_1$ (modified first-order Bessel function of the second kind).

The second condition is thus written:

$$I_1(x_a) + \beta K_1(x_a) = 0$$

Hence:

$$\beta = -\frac{I_1(x_a)}{K_1(x_a)}$$

Series developments of I_0, I_1, K_0 and K_1 may be found in Murray R. Spiegel [SPI 92] and in section 3.3.3. Moreover Goudet [GOU 65] published a table of numerical values.

3.3.3. *Calculation of zero- and first-order modified Bessel functions*

1) Zero order:

$$I_0(x) = 1 + \sum_{r=1}^{\infty} \frac{\left(\frac{x}{2}\right)^{2r}}{(r!)^2}$$

$$K_0(x) = I_0(x)\left[\mathrm{Ln}\frac{x}{2} + C \right] - \sum_{r=1}^{\infty} \frac{\left(\frac{x}{2}\right)^{2r}}{(r!)^2} \left[\frac{1}{r} + \frac{1}{r-1} + \ldots + \frac{1}{1} \right]$$

2) First order:

$$I_1(x) = \frac{x}{2}\left[1 + \sum_{r=1}^{\infty} \frac{\left(\frac{x}{2}\right)^{2r}}{r!(r+1)!}\right]$$

$$K_1(x) = I_1(x)\left[\mathrm{Ln}\frac{x}{2} + C\right] + \frac{1}{x}$$

$$-\frac{x}{4}\sum_{r=1}^{\infty} \frac{\left(\frac{x}{2}\right)^{r2}}{1!(2+1)!}\left[\frac{1}{r+1} + \frac{1}{r} + \frac{1}{r-1} + \ldots + \frac{1}{1} + \frac{1}{r} + \frac{1}{r-1} + \ldots + \frac{1}{1}\right]$$

NOTES.–

Constant C is Euler's constant:

$$C = 0.5772$$

Goudet provides asymptomatic developments of the preceding functions.

3.3.4. *Definition of efficiency*

It is generally assumed, implicitly, that the heat transfer through the cylindrical surface limiting the periphery of each fin is equal to that which would be transferred directly to the surface of the tubes occupied by the fins, if the fins were not present.

Based on this hypothesis, the thermal power Q transferred to the fluid per meter of the tube is the sum of two terms:

1) Direct contact of finless tube with the gas:

$$\alpha\theta_e S_T = \alpha\left(t_e - t_g\right)S_T$$

S_T: external surface of 1 m of bare tube (m^2/m^{-1}).

$$S_T = \pi d_e = 2\pi r_e$$

2) Contact between the gas and the two sides of the fins:

$$N_a \alpha \int_{r_e}^{r_a} (t_m - t_g) dS_a = N_a \alpha 4\pi \int_{r_e}^{r_a} \theta r dr$$

N_a: number of fins per meter of the tube (m^{-1}).

Hence:

$$Q = \alpha (t_e - t_g) S_T + N_a \alpha 4\pi \int_{r_e}^{r_a} \theta r dr$$

The surface of the two sides of a fin is:

$$S_a = 2\pi (r_a^2 - r_e^2)$$

Take:

$$E_a = \frac{4\pi \int_{r_e}^{r_a} \theta r dr}{2\pi (r_a^2 - r_e^2)\theta_e} = \frac{2\int_{r_e}^{r_a} \theta r dr}{(r_a^2 - r_e^2)\theta_e}$$

The power Q transferred per meter of tube can then be written:

$$Q = \alpha \theta_e (S_T + E_a N_a S_a)$$

The coefficient E_a is, by definition, the efficiency of the fins, which modifies the advantage obtained by increasing the surface of the metal in contact with the gas.

3.3.5. *Theoretical efficiency calculation*

We know that:

$$\theta = C\left[I_0(x) + \beta K_0(x)\right]$$

And:

$$x = r\sqrt{\alpha / \lambda e}$$

Where:

$$\int_{r_e}^{r_a} \theta r dr = \frac{\lambda e}{\alpha} \int_{x_e}^{x_a} \theta x dx = \frac{\lambda e C}{\alpha} \left[\int_{x_e}^{x_a} x I_0(x) dx + \beta \int_{x_e}^{x_a} x K_0(x) dx \right]$$

Furthermore (see Goudet):

$$\int x I_0(x) dx = x I_1(x)$$

$$\int x K_0(x) dx = x K_1(x)$$

Hence:

$$\int_{r_e}^{r_a} \theta r dr = \frac{\lambda e C}{\alpha} \left[x_a I_1(x_a) - x_e I_1(x_e) + \beta K_a K_1(x_a) - \beta x_e K_1(x_e) \right]$$

Moreover:

$$x_a I_1(x_a) + \beta x_a K_1(x_a) = x_a \left[I_1(x_a) - \frac{I_1(x_a)}{K_1(x_a)} K_1(x_a) \right] = 0$$

Thus:

$$\int_{r_e}^{r_a} \theta r dr = \frac{-\lambda e C x_e}{\alpha} \left[I_1(x_e) + \beta K_1(x_e) \right]$$

We know that:

$$C = \frac{\theta_e}{I_0(x_e) + \beta K_0(x_e)}$$

Replacing C by this value in the definition of E_a, we obtain:

$$E_a = \frac{-2x_e}{x_a^2 - x_e^2}\left[\frac{I_1(x_e) + \beta K_1(x_e)}{I_0(x_e) + \beta K_0(x_e)}\right]$$

EXAMPLE 3.1.–

$r_e = 0.0125$ m $\qquad r_a = 0.0325$ m

$\alpha = 36$ W.m^{-2}.°C $\qquad \lambda = 55$ W.m^{-1}.°C^{-1}

$e = 0.0005$ m

$$x_e = 0.0125\sqrt{36/(35\times 0.0005)} = 0.452$$

$$x_a = 0.452\times 0.0325/0.0125 = 1.176$$

$I_1\ (1.176) = 0.66$

$K_1\ (1.176) = 0.45$

$I_0\ (0.452) = 1.05$

$I_1\ (0.452) = 0.23$

$K_0\ (0.452) = -1.02$

$K_1\ (0.452) = 1.92$

Hence:

$$\beta = -\frac{0.66}{0.45} = -1.47$$

$$E_a = \frac{-2\times 0.452}{1.176^2 - 0.452^2}\left[\frac{0.23 + (-1.47)\times 1.92}{1.05 + (-1.47)\times(-1.02)}\right]$$

$$E_a = 0.78$$

3.3.6. *Approximate calculation of efficiency*

The method used here was described by Schmidt. According to the author:

$$E_a = \frac{\mathrm{th}(x_e\phi)}{x_e\phi}$$

$$x_e = r_e\sqrt{\alpha/\lambda e} \quad \text{and} \quad \phi = \left[\frac{r_a}{r_e} - 1\right]\left[1 + 0.35\,\mathrm{Ln}\frac{r_a}{r_e}\right]$$

Schmidt demonstrated that, using:

$$0.5 < E_a < 1 \quad \text{and} \quad 1 < r_a / r_e < 8$$

the error is less than 1%.

EXAMPLE 3.2.–

Using the same data as above:

$$x_e = 0.452; \quad \phi = \left(\frac{1.176}{0.452} - 1\right)\left[1 + 0.35\,\mathrm{Ln}\left(\frac{1.176}{0.452}\right)\right]$$

$$x_e\phi = 2.135 \times 0.452 = 0.965$$

$$E_a = \frac{1}{0.965}\left[\frac{e^{0.965} - e^{-0.965}}{e^{0.965} + e^{-0.965}}\right]$$

$$E_a = 0.774$$

This value is very close to the value of 0.78 obtained using the theoretical method.

3.4. Other thermal parameters

3.4.1. *Transfer coefficient: gas side*

The surface occupied by the lateral projection of 1 m of tube is:

$$a_p = d_e + N_a \times 2e \times h$$

d_e: external diameter of the bare tube (m)

N_a: number of fins per meter of the tube (m^{-1})

e: half-thickness of fins (m)

h: fin height (m).

The free area available for the gas to pass through is thus:

$$A_L = A_F - N_{TR} a_p L_T \qquad (A_F \text{ is the frontal surface})$$

N_{TR}: number of tubes per layer: non-dimensional

L_T: tube length (m).

The bulk flow density of the gas is:

$$G = W_G . A_L$$

And the Reynolds number is:

$$Re = \frac{G d_e}{\mu_G}$$

d_e: external diameter of the bare tube (m)

μ_G: viscosity of gas (Pa.s).

Briggs and Young [BRI 63] provide the following formula for the Nusselt number:

$$N_u = \frac{\alpha\, d_e}{\lambda} = 0.134\ Re^{0.681} Pr^{1/3} \left(\frac{s}{h}\right)^{0.2} \left(\frac{s}{e}\right)^{0.113}$$

λ_G: thermal conductivity of the gas ($W/m^{-1}/°C^{-1}$)

Pr: Prandtl number of the gas:

$$Pr = C_P \mu_G / \lambda_G$$

s: spacing between two neighboring fins (m)

(the spacing P_a of the fins is $P_a = s + 2e$).

Hence:

$$\alpha = \frac{\lambda_G Nu}{d_e}$$

3.4.2. *Overall coefficient for bare tubes*

The power transferred per meter of the tube is:

$$Q = \alpha\left(t_e - t_g\right)\left(S_T + E_a N_a S_a\right)$$

The equivalent partial coefficient α_e for bare tubes is such that:

$$Q = \alpha_e\left(t_e - t_g\right) S_T; \qquad S_T = \pi\, d_e$$

Hence:

$$\alpha_e = \alpha\left(1 + E_a N_a \frac{S_a}{S_T}\right); \qquad S_a = \pi\left(h^2 + h d_e\right)$$

Adding the series of resistances, we obtain the overall coefficient U for the bare tubes:

$$\frac{1}{U} = \frac{1}{\alpha_i}\frac{d_e}{d_i} + R_E + \frac{e_T}{\lambda_m} + \frac{1}{\alpha_e}$$

R_E: fouling resistance ($m^2.°C.W^{-1}$) (see Table 2.1)

e_T: tube thickness (m)

α_i: partial coefficient inside the tubes ($W.m^{-2}.°C^{-1}$)

d_e and d_i: external and internal diameters of tubes (m)

λ_m: thermal conductivity of the metal ($W.m^{-1}.°C^{-1}$).

Fouling on the outside of the tubes may be ignored on condition that the system is cleaned periodically using steam jets, compressed air or pressurized water. The situation is different on the process side, but cleaning by mechanical or chemical means remains possible.

3.4.3. *Output temperatures (pure crossflow current)*

Nusselt calculated these temperatures for fluids with non-homogenous temperatures in a given section. We must then take the average value of the temperature in the output section. The following notation will be used:

Number of transfer units, gas side:

$$\theta_G = \frac{US}{W_G C_G}$$

Number of transfer units, procedure side:

$$\theta_P = \frac{US}{W_P C_P}$$

S: external surface of all bare tubes (m^2)

W_G and W_P: bulk flows of gas and process liquid ($kg.s^{-1}$)

C_G and C_P: thermal capacities of gas and process liquid ($J.kg^{-1}.°C^{-1}$).

The reduced temperature on the process liquid side is:

$$T_P = \frac{t_{P2} - t_{G1}}{t_{P1} - t_{G1}}$$

t_{P1} and t_{P2} are the input and output temperatures of the process liquid and t_{G1} is the input temperature of the gas.

Nusselt [NUS 11] provides the following expression:

$$T_P = e^{-\theta}P\left[\psi_o + \psi_1 + \psi_2 + \psi_3\right]$$

$$\psi_o = 1$$

$$\psi_1 = \frac{\theta_P}{1!}\left[1 + \frac{e^{-\theta_G}}{\theta_G} - \frac{1}{\theta_G}\right]$$

$$\psi_2 = \frac{\theta_P^2}{2!}\left[\left(1 + \frac{e^{-\theta_G}}{\theta_G} - \frac{1}{\theta_G}\right) + \left(e^{-\theta_G} + \frac{e^{-\theta_G}}{\theta_G} - \frac{1}{\theta_G}\right)\right]$$

$$\psi_3 = \frac{\theta_P^3}{3!}\left[\left(1 + \frac{e^{-\theta_G}}{\theta_G} - \frac{1}{\theta_G}\right) + \left(e^{-\theta_G} + \frac{e^{-\theta_G}}{\theta_G} - \frac{1}{\theta_G}\right) + \left(\frac{\theta_G e^{-\theta}G}{2} + \frac{e^{-\theta_G}}{\theta_G} - \frac{1}{\theta_G}\right)\right]$$

Hence:

$$t_{P2} = t_{G1} + T_P\left(t_{P1} - t_{G1}\right)$$

The transferred power is:

$$Q_{CR} = W_P C_P\left(t_{P1} - t_{P2}\right)$$

The output temperature of the gas is:

$$t_{G2} = t_{G1} + Q / W_G C_G$$

An average temperature difference for the pure crossflow current can then be calculated using:

$$\Delta T_{CR} = Q_{CR}/US \quad [3.3]$$

3.4.4. *Outlet temperature (pure counterflow current)*

Using the following notation:

$$\theta_P = \frac{US}{W_P C_P} \quad \text{and} \quad R = \frac{W_P C_P}{W_G C_G}$$

The efficiency on the process fluid side is:

$$E_P = \frac{t_{P1} - t_{P2}}{t_{P1} - t_{G1}} = \frac{e^{\theta_P(1-R)} - 1}{e^{\theta_P(1-R)} - R}$$

Hence:

$$t_{P2} = t_{P1} - E_P\left(t_{P1} - t_{G1}\right)$$

The transferred power is:

$$Q_{CE} = W_P C_P\left(t_{P1} - t_{P2}\right)$$

An average temperature difference for a perfect counter-current is:

$$\Delta T_{CE} = Q_{CE}/US \quad [3.4]$$

ΔT_{CE} can be shown to be the logarithmic average of the temperature differences between the hot fluid and the cold fluid at the extremities of the system (see section 5.1.1).

3.4.5. *Crossflow current (multiple passes in relation to gas)*

When all of the layers are combined into a single pass in relation to the gas, we have a pure crossflow current, and a correction coefficient F_1 may be defined, such that:

$$Q_{CR} = F_1 Q_{CE}$$

If the exchanger involves multiple passes in relation to the gas (i.e. N_{PG} passes), coefficient F_N may be obtained using a method set out in the VDI Wärmeatlas:

$$F_N = F_1^{1/N_{PG}}$$

The thermal power is:

$$Q_N = F_N Q_{CE}$$

Hence:

$$Q_N = F_N \Delta T_{CE} US$$

NOTE.–

Bowman *et al.* [BOW 40] provide correction coefficients for crossflow currents in both graphic and analytical forms.

3.5. Determining geometry

3.5.1. *Number of rows and frontal surface*

Given a heating program and the transfer coefficient U, the transfer surface may be estimated using:

$$S = Q/U\Delta T_{CE} \quad [3.5]$$

The face velocity of the gas is given using the following empirical expression:

$$V_F = 5.2 N_R^{-1/3}$$

N_R: number of layers

Or:

$$V_F N_R^{1/3} = 5.2 \quad [3.6]$$

The gas volume flow is known:

$$Q_G = W_F/\rho_G = A_F V_F \quad [3.7]$$

A_F: frontal cross-section: m^2.

The number of tubes per layer is:

$$N_{TR} = (\ell_R/p) - 1 \quad [3.8]$$

p: tube spacing (m)

ℓ_R: width of layer (m).

The heat transfer surface S is equal to the surface of the bare tubes:

$$S = N_T S_T L_T$$

N_T: number of tubes in the system

S_T: external surface of the bare tube per meter of length (m)

L_T: tube length (m).

We know that:

$$L_T = A_F/\ell_R$$

And:

$$N_T = N_{TR} N_R$$

N_{TR}: number of tubes per layer.

Hence:

$$S = \frac{N_{TR}}{\ell_R} N_R S_T A_F$$

Following [3.8]:

$$N_{TR} / \ell_R = \frac{1}{p} - \frac{1}{\ell_R} \# \frac{1}{p}$$

Equation [3.5] gives S and, multiplying equations [3.6], [3.7] and [3.8] member by member, we obtain:

$$V_F N_R^{1/3} \; Q_G N_R A_F = 5.2 \; A_F V_F \; pS/S_T$$

i.e.:

$Q_G N_R^{4/3} = 5.2 \text{ p } S/S_T$ This gives a first estimation of N_R :

$$N_R^{(o)} = \left(\frac{5.2pS}{Q_G S_T} \right)^{3/4}$$

And:

$$A_F^{(o)} = \frac{Sp}{S_T N_R^{(o)}}$$

$N_R^{(o)}$ is not necessarily an integer.

3.5.2. *Mathematical functions*

The two following mathematical functions will be used in the calculations below:

1) "Integer part of x" function:

This function will be noted E(x). For example, if:

$$x = 5.82, \quad \text{then} : E(5.82) = 5$$

2) "Rounded value of x" function:

This function will be noted A(x). More precisely:

$$\text{if } E(x) < x \leq E(x) + 0.5$$

$$\text{then}: A(x) = E(x)$$

$$\text{if } E(x) + 0.5 < x < E(x) + 1$$

$$\text{then}: A(x) = E(x) + 1$$

For example:

$$5 + 0.5 < 5.82 < 6$$

Hence:

$$A(5.82) = 6$$

3.5.3. *Number of tubes per pass*

The number of tubes per pass N_{TP} refers to the number of parallel tubes on the process side. The velocity V_T in these tubes is pre-specified:

– for a liquid:

$$1 \text{ m.s}^{-1} < V_T < 1.5 \text{ m.s}^{-1}$$

– for a condensing vapor, the input speed is of the order of 10 m.s^{-1}.

Hence:

$$N_{TP}^{(o)} = Q_P / V_T a_i$$

$N_{TP}^{(o)}$ is not necessarily an integer.

Q_P: volume flowrate of the process fluid (m^3.s^{-1})

a_i: internal cross-section of the tube (m^2).

However, if the tubes are used to condense pure water vapor, or a pure vapor more generally, the interior of the tubes is at the same temperature everywhere, and the notion of passes on the process ceases to be meaningful. The transfer coefficient for water vapor is:

$$\alpha_i = 7500 \text{ W.m}^{-2} \,°\text{C}^{-1}$$

3.5.4. *Width of tube layers*

In an existing device, this width is:

$$\ell_R = p(N_{TR} + 1)$$

p: transversal spacing of tubes in a layer (m)

N_{TR}: number of tubes in a layer.

When designing a system, the value of ℓ_R must be deduced from knowledge of the frontal surface.

In order for the gas flow to be as uniform as possible, the shape of the frontal section should be as close as possible to a square.

This is particularly important in small-scale devices, which only include a single ventilator. More precisely:

If $A_F < 20 \text{ m}^2$, then the width of a layer is estimated using:

$$\ell_R^{(o)} = \sqrt{A_F}$$

For a larger device, the number of end boxes, i.e. the number of turns, should be kept to a minimum. In this case, longer tubes are used, and the system is equipped with two ventilators. More precisely:

If:

$$20 \text{ m}^2 < A_F < 100 \text{ m}^2$$

write:

$$\ell_R^{(o)} = \sqrt{A_F/2}$$

Finally, if the frontal surface is greater than 100 m^2, the system will be equipped with four ventilators, again in a square configuration, with:

$$\ell_R^{(o)} = \sqrt{A_F}$$

The number of tubes per layer, N_{TR}, is thus:

$$N_{TR}^{(o)} = (\ell_R/p) - 1$$

$N_{TR}^{(o)}$ is not necessarily an integer.

3.5.5. *Number of tubes per layer and per pass*

1) $0.65 < N_{TR}^{(o)}/N_{TP}^{(o)} < 1.5$

Each pass corresponds to a single layer, and we take:

$$N_{TP} = A\left(N_{TP}^{(o)}\right)$$

$$N_{TR} = N_{TP}$$

2) $N_{TR}^{(o)} < 0.65\, N_{TP}^{(o)}$

Each pass corresponds to a number of layers greater than or equal to 2. The number of parallel layers corresponding to a single pass will be:

$$N_{RP} = A\left(N_{TP}^{(o)}/N_{TR}^{(o)}\right) \qquad \text{(A is the rounding function)}$$

The number of tubes per layer will be:

$$N_{TR} = A\left(N_{TP}^{(o)}/N_{RP}\right)$$

The number of tubes per pass is thus:

$$N_{TP} = N_{TR} \times N_{RP}$$

3) $N_{TR}^{(o)} > 1.5\,N_{TP}^{(o)}$

Each layer corresponds to a number of passes greater than or equal to 2. The number of passes per layer will be:

$$N_{PR} = A\left(N_{TR}^{(o)}/N_{TP}^{(o)}\right)$$

The number of tubes per row is thus:

$$N_{TR} = N_{TP} \times N_{PR}$$

Finally, the definitive values of ℓ_R, A_F and L_T are:

The width of a layer, as seen above:

$$\ell_R = p\left(N_{TR} + 1\right)$$

The frontal cross-section:

$$A_F = \frac{Sp}{S_T N_R}$$

The length of the tubes:

$$L_T = A_F/\ell_R$$

3.6. Simulation and direct calculation

3.6.1. *Simulation*

Given the geometry of the system and the operating conditions, it is possible to calculate the counter-current and crossflow current output temperatures, and hence to deduce the corresponding thermal powers Q_{CE} and Q_{CR}.

Thus:

$$F_1 = Q_{CR}/Q_{CE}$$

Given the number of passes N_{PG} in relation to the gas, we deduce:

$$F_N = F_1^{1/N_{PG}}$$

This gives the real theoretical power:

$$Q_N = F_N Q_{CE}$$

t_{P2} and t_{G2} may then be obtained from the heat balances.

In the case of a condenser, the calculation is carried out in sections for the space inside the tubes. For each section, the temperature on the condensation side is considered to be constant and equal to the arithmetic mean of the temperatures at the beginning and end of the section. Given the thermal power for the section, we can deduce the gas output temperature. Each section requires an iterative calculation, as the theoretical potential (MLTD) is dependent on the theoretical power, which is itself dependent on the MLTD.

3.6.2. *Direct calculation*

The intended output temperatures must fulfill certain conditions in order for the problem to be reasonably possible. One sufficient condition is:

$0.2 < R < 1 \qquad \theta_G < 1 - 0.833R^{0.666}$

$1 < R < 4 \qquad \theta_G < 0.7R^{-0.8}$

With:

$$R = W_P C_P / W_G C_G \quad \text{and} \quad \theta_G = US/W_G C_G$$

We therefore know the operating conditions (specifically the power, Q) and the output temperatures. It is then possible to estimate the overall transfer coefficient $U^{(o)}$.

This can be achieved using Table 3.5.

We can then obtain a first estimation of the transfer surface:

$$S^{(o)} = Q/U^{(o)} \Delta T_{CE}$$

Process fluid	$U\left(W.m^{-2}\,°C^{-1}\right)$
Water vapor condensation	800
Water	750
Lightweight organic product	500
Organic vapor condensation	400
Oil	350
Polymers, heavy fuel	130
Gas without phase change	60

Table 3.5. *Overall transfer coefficient*

We then calculate the output temperatures for $S^{(o)}$, the pure crossflow current and the perfect counter-current. From this, we may deduce the corresponding powers and $F_1^{(o)}$, then:

$$F_N^{(o)} = \frac{1}{2}\left(F_1^{(o)} + 1\right)$$

Hence:

$$S^{(1)} = Q/\left(U^{(o)} F_N^{(o)} \Delta T_{CE}\right)$$

It is then possible to carry out a first calculation of the geometry, which provides $U^{(1)}$ and $F_N^{(1)}$, from which we obtain $S^{(2)}$. The process is repeated until S no longer varies.

Condensers always operate based on a single pass on the process side. The simulation calculation is simply carried out by sections, and stopped when the desired temperature is reached; in this way, the necessary length of the tube is obtained immediately.

3.7. Fans

3.7.1. *General points*

Leaving aside ventilators used to blow air from a dryer across an electric or steam air heating system, the ventilators used in air coolers and air condensers are defined by their position, diameter, number and shaft power.

3.7.2. *Position*

Two positions are possible; air always circulates from the bottom to the top in order to benefit from natural draft.

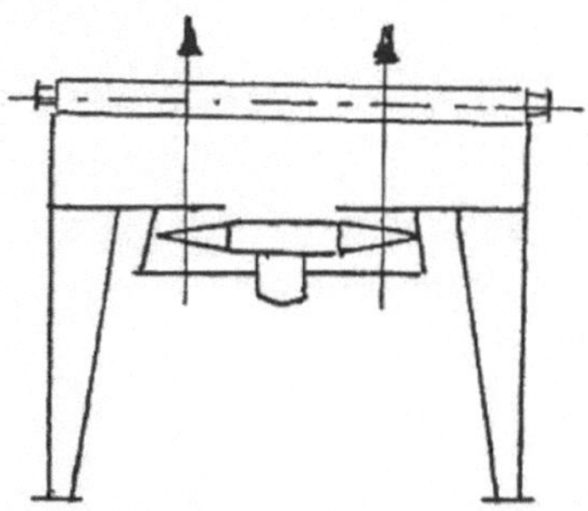

Figure 3.4. *Forced draft*

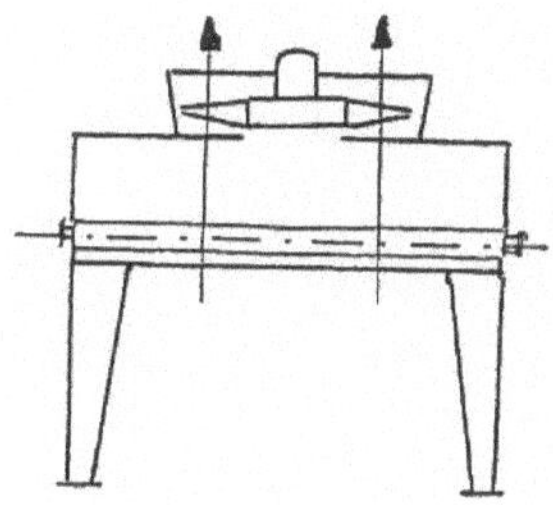

Figure 3.5. *Induced draft*

1) Forced draft (Figure 3.4):

The fan is placed under the tubes.

In this position, the tubes are easy to access. Moreover, the drawing mechanism is not subject to the action of the hot air. Finally, the energy required to blow the air is lower due to the fact that the air is cold.

2) Induced draft (Figure 3.5):

The fan is placed above the tubes.

This configuration is used when working with corrosive fluids with the potential to break through the tubes, in order to eliminate the risk of the process fluid touching the drawing mechanism. The power required to draw the air is higher than in the case of a forced draft. Finally, the distribution of the air is more uniform when drawn than when blown.

3.7.3. *Diameter*

The diameter ϕ_v of a fan cannot be larger than the smallest of the dimensions L_T or ℓ_R. Let this value be ℓ_m.

A large ventilator turning slowly provides a uniform distribution of velocities and produces very little noise, and is also costly (expensive rotor, significant power requirements for startup, etc.). For economic reasons, if ℓ_m is greater than 3 m, the diameter ϕ_v will increase less rapidly than ℓ_m. Thus:

If $\ell_m < 3$ m $\qquad \phi_v = \ell_m$

If $\ell_m \geq 3$ m $\qquad \phi_v = 1.7 + 0.4\ \ell_m$

Finally, due to considerations of mechanical resistance, the diameter ϕ_v is limited to 4.5 m; following the equation above, this corresponds to a value of 7 m for ℓ_m.

3.7.4. *Number*

For the gas velocity to be as uniform as possible across the frontal cross-section, the form of the section should be as close to square as possible:

$$0.5 < L_T/\ell_u < 2$$

However, this constraint can be attenuated by the use of two ventilators rather than one. In this case:

$$2 < L_T/\ell_u < 4$$

or:

$$0,25 < L_T/\ell_u < 0,5$$

Finally, the limitation of the diameter ϕ_v to 4.5 m makes it necessary to install two or four ventilators if ℓ_m is greater than 7 m. However, practical constraints linked to production and installation mean that a value of L_T greater than 10 m is not possible.

3.7.5. *Fan power*

The pressure loss on the gas side can be calculated using a formula developed by Robinson and Briggs [ROB 66]:

$$\Delta P = 18.9 N_R \rho_G V_{GL}^2 Re^{-0.316} \left(\frac{p}{d_e}\right)^{-0.927} \left(\frac{h_{NR}}{p}\right)^{0.515} \quad \text{(Pa)}$$

where:

N_R: number of layers

ρ_G: density of the gas ($kg.m^{-3}$).

$$V_{GL} = Q_G/A_L$$

A_L and Re are defined in the same way as for the calculation of α.

p: tube spacing (m)

d_e: external diameter of tubes (m)

h_{NR}: orthogonal spacing between two layers (m).

For a triangular layout:

$$h_{NR}/p = \sqrt{3/2}$$

The compression ratio τ of the fan is:

$$\tau = P_2/P_1 = (P_1 + \Delta P)/P_1$$

Consequently, the approximate shaft power is:

$$P_a = \frac{W_G}{\eta M_G} RT \, \text{Ln}\tau \qquad (\text{Watt})$$

A figure of 0.65 may be used for the yield η of the machine.

W_G and M_G: bulk gas flow ($kg.s^{-1}$) and molar mass of gas ($kg.kmol^{-1}$)

R and T: perfect gas constant ($J.kmol^{-1}K^{-1}$) and absolute temperature (K).

3.8. Implementation of air coolers

3.8.1. *Regulation*

Regulation techniques may be categorized using two criteria: price and the precision obtained in terms of the temperature of the process fluid output.

In order of price (low-high):

– manually adjustable flaps;

– automatic flaps;

– fan blades with adjustable tilt level (when the fan is stopped);

– self-adjusting fan blades;

– a variable-speed motor.

The precision in terms of process fluid output temperature obtained using vents is ±5°C. Self-adjusting blades and variable speed motors, however, give an acceptable level of precision, in the region of ±1°C.

When a device includes two or more ventilators, multiple systems may be used in combination.

3.8.2. *Frost-proofing*

If the process fluid is water, freezing occurs at 0°C. For liquid petroleum products, however, freezing may occur at higher or lower temperatures.

Whatever fluid is used, the device must be emptied as soon as it stops; for this reason, tubes are always arranged in such a way as to allow drainage by gravity.

If the solidification or flow point of the liquid is higher than 20°C at the minimum ambient temperature (reached on 5 days out of the year), the use of a steam-based air heating device is advisable. This can be achieved by recirculating a portion of the air output through a steam heater.

If the difference between the minimum ambient temperature and the solidification point falls between 10 and 20°C, simple recirculation without steam heating is sufficient.

For differences of less than 10°C, a reduction in the air flow is sufficient to prevent freezing (or frosting).

3.8.3. *Air-side fouling*

Heating engineers do not generally take account of the influence of fouling on fin or tube surfaces; the transfer coefficient, as seen above, is extremely low in relation to the full surface.

In particularly polluted zones, however, fouling may result in an increase in air pressure loss as the air moves through the tube bundle. In these circumstances, a square tube layout should be used and the space left between the tubes should be increased.

4

Scraped Surface Heat Exchangers, Spherulation Towers and Solidification on a Moving Surface

4.1. General points

4.1.1. *Benefits when using viscous products*

Falling film mechanisms can only be used if the viscosity of the liquid in the tube remains below 200 centipoises, and if the fraction of the liquid which is vaporized does not exceed a value of between 20 and 40% of the input flow by mass. If these conditions are not respected, the film will no longer form in a satisfactory manner (dry zones) and partial recycling becomes necessary.

Devices used for highly viscous products are cylindrical in shape, and the double-wall cavity is filled with an utility fluid, which may be a heat transfer liquid (for heating or cooling) or a frigorific fluid (for boiling liquids). The body of the apparatus is not completely filled with the product, and continuous movement is ensured by the rotation of a shaft running the length of the column axis, equipped with either a screw mechanism or inclined blades which continually scrape down the wall.

The portion of product in contact with any given part of the wall therefore changes with each passage of the sweeping or scraping system.

Figure 4.1. *Scraping*

4.1.2. *Possible designs and heat transfer*

Between two scrapings, heat energy is transmitted to the fluid by pure conduction, and the viscosity of the product is no longer relevant. However, this is strictly true only if the rotor completely eliminates the product at the wall; in principle, this would require there to be no gap whatsoever between the rotor and the wall.

In reality, this situation is impossible, and feasible devices may be grouped into three categories:

1) quasi-gapless mechanisms (gap between 0.1 and 1 cm) created by precision machining;

2) non-machined rotor mechanisms (using plastic or wooden scrapers), which have a gap of between 0.1 and 1 mm, depending on wear;

3) wide gap mechanisms (gap measured in cm). In this case, high product viscosity is an advantage, as it can be pulled (rather than pushed) by the rotor.

The way in which heat transfer is evaluated depends on the type of design, particularly with regard to the partial coefficient on the product side α_P. The overall coefficient U is such that:

$$\frac{1}{U}=\frac{1}{\alpha_P}+\frac{e_p}{\lambda_p}+\frac{1}{\alpha_S}$$

e_p: wall thickness (m)

λ_p: thermal conductivity of the wall ($W.m^{-1}.°C^{-1}$)

α_S: partial coefficient, utility fluid side ($W.m^{-2}.°C^{-1}$).

The table below provides estimations of α_S.

Type of operating fluid	α_S ($W.m^{-2}.°C^{-1}$)
water	5,200
water vapor	14,000
brine	4,600
boiling liquid	9,300

Table 4.1. *Operating fluid coefficients*

A method for estimating α_P will be described for each of the three device types described above.

Note that, on the product side, the wall temperature t_p is:

$$t_p = \frac{1}{\alpha_P}\left[Ut_S + (\alpha_P - U)t_P\right]$$

where t_S and t_P are the temperatures of the utility fluid and the product, respectively.

When heating an organic product, the wall temperature t_p must be limited in order to avoid "cooking" the product, which can result in baked-on deposits forming on the wall, particularly if the spacing between the wall and the rotor is relatively large. An upper limit of between 130 and 160°C is often used.

4.2. Types of scraped surface heat exchanger

4.2.1. *Quasi-gapless mechanisms, almost completely immune to wear*

For reasons of practicality during production, these exchangers are built on a modest scale, and all share the following characteristics:

Internal diameter	0.3–0.6 m
Rotation speed	100–400 $rev.mn^{-1}$
Overall heat transfer coefficient	700–2,300 $W.m^{-2}.°C^{-1}$

The rotor includes a scraper in the form of a single- or double-thread screw. The rotation speed is relatively low (200 rev.mn^{-1}) for particularly viscous products, such as ground meat, but set at maximum (400 rev.mn^{-1}) for less viscous products, such as pea purée, in order to ensure that the product flow remains regular.

The shell and rotor are made of the same material (noble metal) so that the gap remains constant, independent of the temperature.

We will show that, the average flow density entering a wall during time T is:

$$\varphi_m = \frac{2}{\sqrt{\pi}}\sqrt{\frac{\lambda_P \rho_P C_P}{T}}\left(t_p - t_P\right)$$

The heat propagation equation for the wall is:

$$\frac{\partial t}{\partial \tau} = \frac{\lambda}{\rho\, C_p}\frac{\partial^2 t}{\partial x^2}$$

with the following boundary conditions between time τ and temperature t:

for $\begin{vmatrix} \tau < 0 \\ \forall x \end{vmatrix}$ $t = t_0$

for $\begin{vmatrix} \tau \geq 0 \\ x = 0 \end{vmatrix}$ $t = t_1$

This differential equation has a classic solution:

$$\frac{t_p - t}{t_p - t_P} = \operatorname{erf}(u) = \frac{2}{\sqrt{\pi}}\int_0^u e^{-u2} du \text{ with } u = \frac{x}{2\sqrt{\dfrac{\lambda T}{\rho C_p}}}$$

x: depth in the wall (m)

λ: conductivity of wall (W.m^{-1}.°C^{-1})

ρ: density of wall ($kg.m^{-3}$)

C_p: specific heat of the wall ($J.kg^{-1}.°C^{-1}$).

The instantaneous heat flow density entering the wall is:

$$\varphi_I = -\lambda\left[\frac{\partial t}{\partial x}\right]_{x=0} = \lambda\left(t_p - t_P\right)\left[\frac{\partial \mathrm{erf}(u)}{\partial u}\right]_{x=0}\left[\frac{\partial u}{\partial x}\right]_{x=0}$$

$$\varphi_I = \frac{1}{\sqrt{\pi}}\sqrt{\frac{\lambda\rho C_P}{\tau}}\left(t_p - t_P\right)$$

The average heat flow density during time T is:

$$\varphi_m = \frac{1}{T}\int_0^T \varphi_I d\tau = \frac{1}{T}\frac{\sqrt{\lambda\rho C_P}}{\sqrt{\pi}}\left(t_p - t_P\right)\int_0^T \frac{d\tau}{\sqrt{\tau}}$$

$$\varphi_m = \frac{2}{\sqrt{\pi}}\sqrt{\frac{\lambda\rho C_P}{T}}\left(t_p - t_P\right)$$

φ_m is given in $W.m^{-2}$.

Indices P and p characterize the product and the wall, respectively. With N and n being the rotor rotation frequency and the number of threads in the screw, the scraping period is:

$$T = \frac{1}{Nn} \qquad \left(T \text{ in s and N in } s^{-1}\right)$$

Hence, the expression of coefficient α_P (with $2/\sqrt{\pi} = 1.13$):

$$\alpha_P = 1.13 e_\alpha \sqrt{nN\lambda_P\rho_P C_P}$$

where e_α may be referred to as the *thermal efficiency of the scraper* which, in this case, is around 0.75.

When evaluating the thermal properties of highly varied and complex products, such as tomato juice, ground meat, jelly or mayonnaise, we simply consider the product as a mixture of water and an average organic component:

$$\lambda_P = 0.6x + 0.12(1-x) \qquad \left(W.m^{-1}.°C^{-1}\right)$$

$$C_P = 4180\ x + 2100(1-x) \qquad \left(J.kg^{-1}.°C^{-1}\right)$$

x: mass fraction, of water in the product (total humidity).

The expression of α_P shows that this coefficient increases at the same rate as the square root of the rotation frequency; this is why the overall coefficient will virtually cease to increase above a peripheral scraping speed of 10 $m.s^{-1}$, whilst limiting undesirable wear on the rotor.

EXAMPLE 4.1.–

A food product with a water content of 70% is to be heated using steam. We need to find the heat transfer coefficient:

ρ_P : 950 $kg.m^{-3}$ $\qquad$ $e_p = 3\times 10^{-3}$ m

$N = 200$ $rev.mn^{-1} = 3.33\ s^{-1}$ $\qquad$ $\lambda_p = 15\ W.m^{-1}°C^{-1}$

$x = 0.7$ $\qquad$ $e_\alpha = 0.75$

$n = 2$

$$C_P = 4180\times 0.7 + 2100\times 0.3 = 3556\ J.kg^{-1}.°C^{-1}$$

$$\lambda_P = 0.6\times 0.7 + 0.12\times 0.3 = 0.456\ W.m^{-1}.°C^{-1}$$

$$\alpha_P = 1.13\times 0.75\sqrt{2\times 3.33\times 0.456\times 3556} = 2720\ W.m^{-2}.°C^{-1}$$

$$\frac{1}{U} = \frac{1}{14000} + \frac{3.10^{-3}}{15} + \frac{1}{2720}$$

$$U = 1564\ W.m^{-2}.C^{-1}$$

4.2.2. *Mechanisms with small gap size, subject to wear*

In the wine industry, devices with plastic or wooden scraper blades are used to crystallize tartrates. These blades have a low thermal efficiency of around 0.35, compared with 0.75 in the case of metal blades.

The investment required per square meter is considerably lower, but the surface requirement is practically doubled.

4.2.3. *Mechanisms with a wide gap between the blade and the wall*

For certain products, a space of around a fraction of centimeter between the wall and the screw or scraper blades is needed in order to avoid blockages due to hard particles becoming jammed between the rotor and the wall.

This is the case for:

– pre-dryers for biological slurries;

– cooling screws with a double-walled trough for cooling divided solids exiting a kiln or dryer;

– screw crystallizers with a double-walled cooling system (used to crystallize paraffin products in the oil industry).

The coefficient on the product side is calculated differently in this case, as it involves divided solids or highly viscous products, with an effective viscosity of between 1 and 100 Pa.s. depending on humidity.

Let s be the dryness of a product, i.e. the mass fraction of anhydrate components in the wet product. Thus:

$$\alpha_P = 700 - 600\ s \qquad \left(W.m^{-1}.°C^{-1}\right)$$

The example below concerns the estimation of the surface of a pre-dryer for biological slurries.

EXAMPLE 4.2.–

Pre-drying of food slurry at 60°C:

Flow of dry matter: $W_s = 0.17$ kg.s^{-1} $C_s = 1{,}250$ J.ks^{-1}.°C^{-1}

Flow of humidifying water: $W_e = 0.83$ kg.s^{-1} $C_e = 4{,}180$ J.kg^{-1}.°C^{-1}

0.49 kg.s^{-1} of water is vaporized, as the boiling delay is negligible. The utility fluid is steam at a temperature of 160°C (5 bar G). The wall is made of stainless steel with a thickness of 5 mm (λ_p = 15 W.m^{-1}.°C^{-1}). The simplified equivalent exchanger method will be used.

1) Pre-heating:

$$(100-60)(0.17\times 1250+0.83\times 4180)=147\,276 \text{ W}$$

During this operation, the dryness does not change:

$$s_{pr}=\frac{0.17}{0.17+0.83}=0.17$$

$$\alpha_p=700-600\times 0.17=598 \text{ W.m}^{-2}.°\text{C}^{-1}$$

$$\frac{1}{U_{pr}}=\frac{1}{598}+\frac{0.005}{15}+\frac{1}{14000}$$

$$U_{pr}=481 \text{ W.m}^{-2}.°\text{C}^{-1}$$

$$\text{DLMT}=\frac{100-60}{\text{Ln}\dfrac{100}{60}}=78.3°\text{C}$$

$$S_{pr}=\frac{147\,276}{78.3\times 481}=3.9 \text{ m}^2$$

2) Vaporization:

$$0.49 \times 2.257 \times 10^6 = 1.106 \times 10^6 \text{ W}$$

$$\text{Average dryness: } \bar{s} = \frac{1}{2}\left(0.17 + \frac{0.17}{1-0.49}\right) = 0.25$$

$$\alpha_P = 700 - 600 \times 0.25 = 550 \text{ W.m}^{-2}.°\text{C}^{-1}$$

$$\frac{1}{U_{va}} = \frac{1}{550} + \frac{0.005}{15} + \frac{1}{14000}$$

$$U_{va} = 450 \text{ W.m}^{-2}.°\text{C}^{-1}$$

$$\text{DLMT} = 160 - 100 = 60°\text{C}$$

$$S_{va} = \frac{1.106 \times 10^6}{60 \times 450} = 41 \text{ m}^2$$

Giving the total surface S_T of the cylindrical heating wall:

$$S_T = 3.9 + 41 \# 45 \text{ m}^2$$

4.3. Spherulation towers

4.3.1. *Introduction*

Spherulation is solidification that occurs following the atomization of a liquid, whereby droplets, cooled in an air current, are transformed into small solid spheres, or spherules.

This process is mostly applied to ammonium nitrate and to urea, both of which are fertilizers. Historically, lead shot was made in a similar way, by spraying molten lead into water.

Spherulation systems are costly to establish but relatively inexpensive to use, and are suitable for products with a low crystallization temperature, as the thermal capacity of the air is low. If the solidification point of the product is also high, the state change will also result in a high thermal potential (temperature difference between the product and the air).

After leaving the tower, spherules are subjected to an additional drying process, as they still retain a certain percentage of humidity. They are then cooled and sorted, and excessively small or grainy spherules are recycled.

4.3.2. *Droplet formation*

Pressurized liquid is generally directed through a nozzle, and the jet then disperses into droplets. The diameter of these droplets is around twice that of the orifice, with very uniform grain size. However, this atomization also results in the creation of "mini droplets" with a much smaller size than the spherules; these are carried off by the air, creating an atmospheric pollution issue.

Nozzle-based spraying techniques are suitable for liquids with a viscosity lower than 100 Pa.s, such as urea or ammonium nitrate. A turbine or disc mechanism may also be used, as in the case of spray drying towers. These techniques produce less uniform grains.

4.3.3. *Practical data*

The molten product is atomized at a temperature of 10–20°C above its solidification point. Use of a higher temperature would increase the thermal power to evacuate without any benefits; a lower temperature would increase the risk of early crystallization, blocking the tubes.

The air at the top of the tower needs to be below 60°C to enable users to access the atomization system without any risk of injury.

Droplets rapidly attain their maximum falling velocity in the upward air current. Practically, a tower processing 500,000 tonnes of product per year will have a diameter of 15 m, and the freefall height for the product will be around 60 m.

The total height of the device may reach 100 m, if the bottom of the tower includes a funnel to collect the product and distribute it onto a transport system. A funnel of this type may have a height of 20 m. Savings may be made in terms of total height by using a flat base with rakes which scrape the product onto a transporter for removal.

To prevent the product from sticking to the wall, the wall may be coated in rubber or a plastic material; some form of wall protection is always required, as the main structure is made of concrete.

4.3.4. *Dimensioning*

The tower dimensioning process involves the following operations:

1) Overall heat balance and enthalpic graph

Given the input and output temperatures of the air, this balance enables us to identify the air flow. The thermal power is obtained by taking account of the sensible and latent heat, along with the heat resulting from transitions between crystalline forms, where applicable.

We then trace a type of enthalpic graph similar to that used when designing condensers, and define sections of thermal power ΔQ_i. The height of each section must then be determined. The power values ΔQ_i are not necessarily equal.

The procedure for a single section is given below. For each section, the physical properties are defined at the arithmetic mean of the temperatures at the section boundaries.

2) Maximum fall velocity of droplets and spherules

The following method may be used.

Calculate:

$$X = \frac{4gd_p^3(\rho_S - \rho_G)}{3\mu_G^2}$$

d_p: diameter of droplets or spherules (m)

g: acceleration due to gravity (9.81 m.s^{-2})

μ_G: viscosity of gas (Pa.s)

ρ_S and ρ_G: bulk densities of solid and gas (kg.m^{-3}).

Y is then deduced using:

$1 < X < 10^4$ $$Y = \frac{13842}{X^{1.57}} + \frac{218}{X^{1.074}}$$

$10^4 < X < 10^9$ $$Y = \frac{59.11}{X^{0.933}} + \frac{0.0278}{X^{0.378}}$$

The maximum fall velocity can then be obtained directly:

$$V_\ell = \left[\frac{4g\mu_G(\rho_S - \rho_G)}{3\rho_G{}^2 Y}\right]^{1/3}$$

EXAMPLE 4.3.–

$d_p = 0.3 \times 10^{-3}$ m $\qquad \mu_G = 20 \times 10^{-6}$ Pa.s

$\rho_S = 1{,}725$ kg.m^{-3} $\qquad \rho_G = 1.20$ kg.m^{-3}

$$X = \frac{4 \times 9.81 \times (0.3)^3 \times 10^{-9} \times 1725}{3 \times \left(20 \times 10^{-6}\right)^2} = 1523$$

$$Y = \frac{13842}{1523^{1.57}} + \frac{218}{1523^{1.074}} = 0.2226$$

$$V_\ell = \left[\frac{4 \times 9.81 \times 20 \times 10^{-6} \times 1725}{3 \times 1.2^2 \times 0.2226}\right]^{1/3}$$

$$V_\ell = 1.12 \text{ m.s}^{-1}$$

3) Heat transfer coefficient:

The Nusselt number is obtained using:

$$Nu = 2 + 0.6\ Re^{0.5} Pr^{0.33}$$

The resulting transfer coefficient is:

$$\alpha = \frac{\lambda Nu}{d_p}$$

In these expressions:

Re is the Reynolds number:

$$Re = \frac{V_\ell d_p \rho_G}{\mu_G}$$

Pr is the Prandtl number:

$$Pr = \frac{C_P \mu_G}{\lambda_G} \qquad (0.7 \text{ for air})$$

C_P: thermal capacity of gas: (1,010 J.kg^{-1}.°C^{-1} for air)

μ_G: viscosity of gas (Pa.s)

λ_G: thermal conductivity of gas (W.m^{-1}.°C^{-1})

EXAMPLE 4.4.–

$V_\ell = 1.12$ m.s^{-1} $\quad d_p = 0.3 \times 10^{-3}$ m $\quad \rho_G = 1.2$ kg.m^{-3}

$\mu_G = 20 \times 10^{-6}$ Pa.s $\quad Pr = 0.7 \quad \lambda_G = 0.027$ W.m^{-1}.°C^{-1}

$$Re = \frac{1.12 \times 0.3 \times 10^{-3} \times 1.2}{20 \times 10^{-6}} = 20.16$$

$$Nu = 2 + 0.6 \times (20.16)^{0.5} \times 0.7^{0.33} = 4.4$$

$$\alpha = \frac{0.027 \times 4.4}{0.3 \times 10^{-3}} = 396 \text{ W.m}^{-2}.\text{°C}^{-1}$$

4) Section height:

With Q_G the volumetric flow rate of the gas and A_T the cross-section of the tower, the gas velocity is $V_G = Q_G/A_T$.

The fall velocity of the particles in relation to the device is:

$$V_\ell - V_G$$

and the time spent by particles in a section of height ΔH is:

$$\Delta\tau = \frac{\Delta H}{V_\ell - V_G}$$

The mass flow of particles of diameter d_p is W_p, and their true density is ρ_p; therefore the volume flow is $Q_p = W_p/\rho_p$.

The surface area of a sphere is $6/d_p$, meaning that the surface of the particles present in a section is:

$$\Delta S = Q_p \ \Delta\tau \times \frac{6}{d_p} = \frac{W_p \Delta H \times 6}{\rho_p (V_\ell - V_G) d_p}$$

Taking ΔT to denote the MLTD (mean logarithmic temperature difference) in the section and given the thermal power ΔQ of the section, we can write:

$$\Delta Q = \alpha \Delta S \Delta T$$

Replacing factor ΔS by its value, we obtain:

$$\Delta H = \frac{\Delta Q \rho_p (V_\ell - V_G) d_p}{6 \alpha \Delta T W_p}$$

NOTE.–

V_ℓ and α are calculated using the highest value of d_p produced by the atomization system. This is to ensure that even the largest particles will be solid on reaching the bottom of the tower, otherwise the product will be difficult, if not impossible, to remove.

4.3.5. *Pollution*

Spherulation towers expel an average of 10,000 m^3/h of air per tonne of product. This air contains an average of 50 mg/m^3 of mini-spherules and dust particles of less than 5 μm in size. The dust content decreases rapidly if the speed of the air in the tower is reduced from 3 to 1 $m.s^{-1}$.

Particles with a size of less than 1 μm result in the production of a plume above the tower. This can be avoided by thorough wet cleaning of the air, or by simply recycling this gas.

4.4. Continuous solidification on a moving surface

4.4.1. *Principle*

A molten product flows onto a cooled surface which is moving at constant speed. The product is retrieved at the end of the conveyor, ready to be packaged.

The surface carrying the product may be a thin metal plate, sliding over a steel deck and cooled using water circulation.

This surface may also take the form of a cylinder, turning on a horizontal axis. In this case, water circulating within the double wall of the cylinder is used for cooling purposes.

4.4.2. *Elements involved in a calculation*

This process includes three steps:

– cooling of the liquid;

– solidification;

– cooling of the solid.

The product, whatever its state, is never at the same temperature right the way through; a finite element method may therefore be used for precise calculations.

The heat penetration method may also be used (see section 4.2.1).

Using a conveyor system, the coolant circulates through the deck in the opposite direction to the product (i.e. countercurrent). In the case of a cylinder, the coolant circulates parallel to the axis of the cylinder, i.e. with a crossflow current in relation to the product.

If the metallic wall between the product and the coolant is thick, then the thermal capacity of the wall should be taken into consideration, as its temperature will no longer be homogeneous throughout.

We shall not go into further detail concerning these devices here, as they are rarely used in industrial processes.

5

Efficiency of Single-phase Heat Exchangers: Fouling

5.1. Number of transfer units and efficiency

5.1.1. *Mean logarithmic temperature difference (MLTD)*

The heat Q transferred in an exchanger may be shown to be equal to:

$$Q = U\,S\,\Delta T$$

With:

U: overall exchange coefficient ($W.m^{-2}.°C^{-1}$)

S: transfer surface (m^2)

ΔT: mean logarithmic temperature difference (MLTD) (°C)

Q: transferred heat (W).

We shall use the example of a perfect counter-current exchanger.

Figure 5.1 shows variations in the temperature of the hot and cold fluids along the length of the exchange surface:

T: temperature of hot fluid

t: temperature of cold fluid.

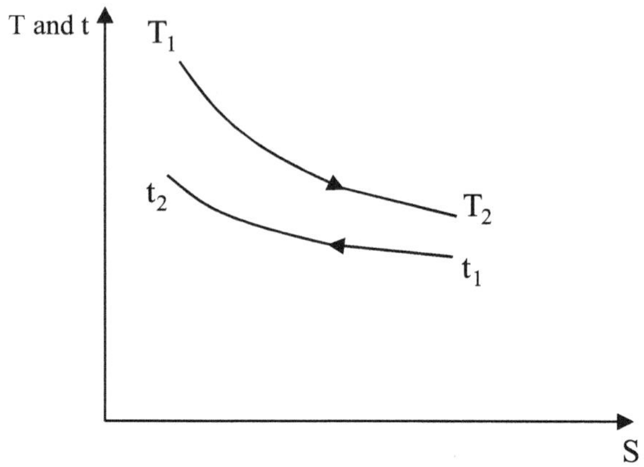

Figure 5.1. *Evolution of temperatures using counter-current*

A positive increase dS in the surface corresponds to a reduction in temperatures T and t. We may therefore write:

$$dQ = UdS(T - t) = -W\,C\,d\,T = -wcdt$$

W and w: mass flowrates of hot and cold fluid ($kg.s^{-1}$)

C and C: thermal capacities of hot and cold fluid ($J.kg^{-1}.°C^{-1}$)

Or:

$$UdS(T - t) = -\frac{d\,T}{1/WC} = \frac{dt}{-1/wc} = -\frac{d(T - t)}{1/WC - 1/wc}$$

Taking R = WC/wc, we obtain:

$$\frac{UdS}{WC}(1 - R) = -\frac{d(T - t)}{T - t}$$

Integrating:

$$(1-R)\frac{US}{WC}=-Ln\frac{(T_2-t_1)}{(T_1-t_2)}=Ln\frac{(T_1-t_2)}{(T_2-t_1)} \qquad [5.1]$$

The exchanged heat may also be written:

$$Q=WC(T_1-T_2)=US\,\Delta T$$

i.e.:

$$T_1-T_2=\frac{US}{WC}\Delta T \qquad [5.2]$$

By eliminating US/WC in equations [5.1] and [5.2]:

$$\Delta T=\frac{(T_1-T_2)(1-R)}{Ln\frac{(T_1-t_2)}{(T_2-t_1)}}$$

And:

$$R=\frac{t_2-t_1}{T_1-T_2}$$

Thus:

$$(T_1-T_2)(1-R)=T_1-T_2-(t_2-t_1)=(T_1-t_2)-(T_2-t_1)$$

Hence, the Mean Logarithmic Temperature Difference:

$$MLTD=\Delta T=\frac{(T_1-t_2)-(T_2-t_1)}{Ln\frac{T_1-t_2}{T_2-t_1}}$$

Similarly, in the case of a co-current exchange, the fundamental formula remains valid:

$$Q = U\,S\,\Delta\,T$$

NOTES.–

1) In the case of a counter-current transfer, if $WC = wc$, i.e. if $R = 1$, then:

$$dS = WC\,dT = w\,c\,d\,t$$

i.e.:

$$WC\,d(T - t) = 0 \text{ and } T - t = \text{cste}$$

In this case, ΔT is the difference in temperature from one end of the exchanger to the other.

2) In the case of a tubular exchanger with multiple passes on the tube side, the MLTD must be corrected using a correction factor F, as we shall see. The fundamental formula becomes:

$$Q = U\,S\,F\,\Delta\,T$$

Bowman *et al.* [BOW 40] provide expressions of F for different exchanger configurations.

5.1.2. *Number of transfer units (NTU)*

The number of transfer units is the relationship:

$$\Theta = \frac{T_1 - T_2}{F\,\Delta\,T} \quad \text{or} \quad \theta = \frac{t_2 - t_1}{F\,\Delta\,T}$$

As we have seen, we may write:

$$Q = U\,S\,F\,\Delta\,T = WC(T_1 - T_2) = wc(t_2 - t_1)$$

Hence:

$$\Theta = \frac{US}{WC} \quad \text{and} \quad \theta = \frac{US}{wc}$$

The two numbers of transfer units are therefore linked by the relationship:

$$\frac{\Theta}{\theta} = \frac{wc}{WC} = \frac{1}{R}$$

5.1.3. *Efficiency*

1) Definitions:

The efficiency of an exchanger may be defined as the relationship between the quantity of heat which is actually transferred and the quantity of heat which could have been transferred.

The first heat quantity is proportional to:

$$T_1 - T_2 \quad \text{or} \quad t_2 - t_1$$

depending on which of the two fluids is chosen.

The second heat quantity is proportional to $T_1 - t_1$.

The efficiency is thus written:

– for the hot fluid:

$$E = \frac{T_1 - T_2}{T_1 - t_1}$$

– for the cold fluid:

$$e = \frac{t_2 - t_1}{T_1 - t_1}$$

2) Relationships between E, R and Θ in the case of single-pass counter-current exchangers:

We have seen that:

$$\Theta(1-R)=\mathrm{Ln}\frac{(T_1-t_2)}{T_2-t_1}$$

Furthermore:

$$T_1-t_2=T_1-t_1-(t_2-t_1)=(T_1-t_1)(1-RE)$$

$$T_2-t_1=T_1-t_1-(T_1-T_2)=(T_1-t_1)(1-E)$$

Thus:

$$\Theta=\frac{\mathrm{Ln}\left(\dfrac{1-RE}{1-E}\right)}{1-R}$$

Similarly:

$$\theta=\frac{\mathrm{Ln}\left(\dfrac{1-er}{1-e}\right)}{1-r}$$

Inversely, we may write:

$$E=\frac{\exp(1-R)\Theta-1}{\exp(1-R)\Theta-R}$$

$$e=\frac{\exp(1-r)\theta-1}{\exp(1-r)\theta-r}$$

When R = 1, the expression of Θ becomes indeterminate. It may be evaluated by taking the derivative of the numerator and the denominator in relation to R. We obtain:

$$\theta=e/(1-e) \quad \text{and} \quad e=\theta/(1+\theta)$$

3) In the case of a co-current exchange, the efficiency is always defined as:

$$E = \frac{T_1 - T_2}{T_1 - t_1} \quad \text{and} \quad e = \frac{t_2 - t_1}{T_1 - t_1}$$

$$R(T_1 - T_2) = t_2 - t_1$$

$$\Delta T_1 = (T_1 - t_1)$$

$$\Delta T_2 = T_1 - t_1 - \left[(T_1 - T_2) + (t_2 - t_1)\right] = T_2 - t_2$$

$$= T_1 - t_1 - (1+R)(T_1 - T_2)$$

$$= T_1 - t_1 - (1+R)E(T_1 - t_1)$$

$$= (T_1 - t_1)\left[1 - E(1+R)\right]$$

$$\frac{\Delta T_1}{\Delta T_2} = \frac{1}{1 - E(1+R)}$$

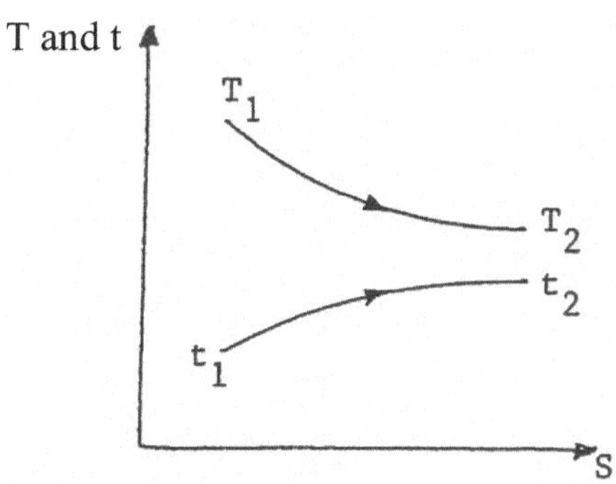

Figure 5.2. *Co-current transfer*

Moreover:

$$U.dS\left(T-t\right)=dQ=\frac{-dT}{1/WC}=\frac{+dt}{1/WC}=\frac{-d\left(T-t\right)}{\dfrac{1}{WC}+\dfrac{1}{wc}}$$

However, R = WC/wc, so,

$$\frac{U\times dS}{WC}\left(1+R\right)=\frac{-d\left(T-t\right)}{T-t}$$

$$\left(1+R\right)\frac{US}{WC}=Ln\left[\frac{T_1-t_1}{T_2-t_2}\right]=Ln\frac{\Delta T_1}{\Delta T_2}$$

$$\Theta=\frac{Ln\left[\dfrac{1}{1-E\left(1+R\right)}\right]}{1+R}\text{ and }\theta=\frac{Ln\left[\dfrac{1}{1-e\left(1+r\right)}\right]}{1+r}$$

$$\exp\left(-\Theta\left(1+R\right)\right)=1-E\left(1+R\right)$$

$$E=\frac{1-\exp\left(-\Theta\left(1+R\right)\right)}{1+R}\text{ and }e=\frac{1-\exp\left(-\theta\left(1+r\right)\right)}{1+r}$$

5.1.4. *Applications of the notion of efficiency*

1) Determination of exchanger output temperatures.

Take a co-current exchanger, characterized by:

$U = 174\ W.m^{-2}.°C^{-1}$

$S = 40\ m^2$

Hot fluid	Cold fluid
$W = 1.667\ kg.s^{-1}$	$w = 2.778\ kg.s^{-1}$
$C = 2{,}090\ J.kg^{-1}.°C^{-1}$	$c = 2{,}508\ J.kg^{-1}.°C^{-1}$
$T_1 = 316°C$	$t_1 = 38°C$
$T_2 = ?$	$t_2 = ?$

$$R = WC/wc = 1.66 \times 2090/(2.778 \times 2508) = 0.5$$

$$\Theta = US/WC = 174 \times 40/(1.0667 \times 2090) = 2$$

$$E = \frac{1 - \exp(-2 \times 1.5)}{1{,}5} = 0.633$$

$$E = (T_1 - T_2)(T_1 - t_1)$$

i.e.

$$0.633 = (316 - T_2)/(316 - 38)$$

$$T_2 = 140°C$$

$$R = (t_2 - t_1)/(T_1 - T_2)$$

$$0.5 = (t_2 - 38)/(316 - 140)$$

$$t_2 = 126°C$$

2) Determination of intermediate temperatures in a series of exchangers.

Take the example of a system with two exchangers.

Based on the considerations above, the efficiencies of the two exchangers are known.

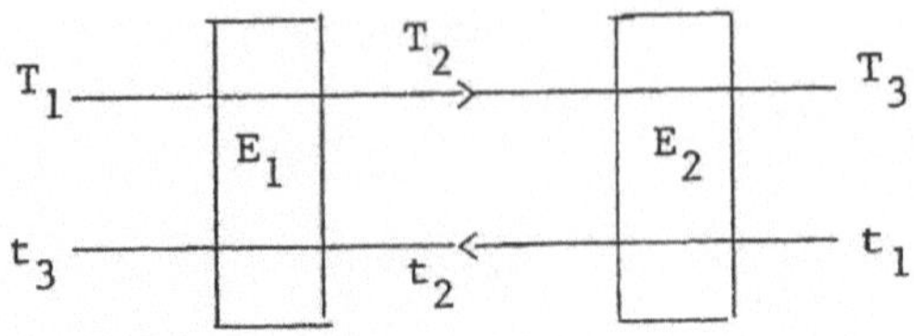

Figure 5.3. *Exchangers placed in series*

We can therefore write:

$$E_2 = \frac{T_2 - T_3}{T_2 - t_1}$$

In this expression, the only unknown quantity is T_2; the value of T_2 can thus be obtained from the expression.

Now, write:

$$E_1 = \frac{T_1 - T_2}{T_1 - t_2}$$

In this expression, the only unknown quantity is t_2, and this temperature can therefore be calculated.

3) Overall efficiency of exchangers in a series.

Co-current fluids:

$$E = \frac{1 - \prod_{1}^{n}\left[1-(1-R)E_i\right]}{1+R}$$

Counter-current fluids:

$$1-E=\frac{1-R}{\prod_{1}^{n}\left[\frac{1-RE_i}{1-E_i}\right]-R}$$

4) Thermal power Q of an *intermediate loop* system:

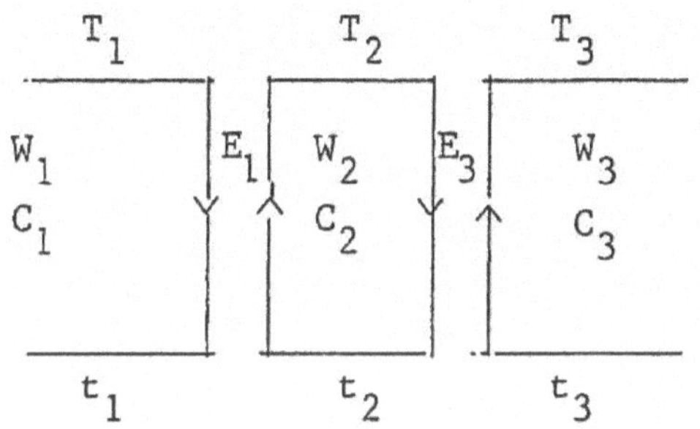

Figure 5.4. *Intermediate loop*

We can write:

$$E_1=\frac{T_1-t_1}{T_1-t_2}=\frac{Q/W_1C_1}{T_1-t_2} \quad \text{hence: } T_1-t_2=\frac{Q}{E_1W_1C_1}$$

$$E_3=\frac{T_3-t_3}{T_2-t_3}=\frac{Q/W_3C_3}{T_2-T_3} \quad \text{hence: } T_2-t_3=\frac{Q}{E_3W_3C_3}$$

Now, adding member by member:

$$T_1-t_3+T_2-t_2=Q\left[\frac{1}{E_1W_1C_1}+\frac{1}{E_3W_3C_3}\right]$$

And:

$$T_2-t_2=\frac{Q}{W_2C_2}$$

Consequently:

$$Q = \frac{T_1 - t_3}{\frac{1}{E_1 W_1 C_1} + \frac{1}{E_3 W_3 C_3} - \frac{1}{W_2 C_2}}$$

In this way, we can obtain the thermal power exchanged as a function of the input temperatures of the extreme fluids. Efficiencies are defined in relation to the surfaces and coefficients of the two exchangers.

Intermediate loop systems are used for cooling in nuclear power plants. The radiation should not be allowed to reach the cooling water.

5.1.5. *Multipass exchangers and coefficient F*

An exchanger can have multiple passes on the side of the tube and one passe on the shell side. In this case, the MLTD must be corrected by a factor F and the transferred power is:

$$Q = U\,S\,F\,\Delta\,T$$

In accordance with the previous statements, we can write:

$$e = \frac{t_2 - t_1}{T_1 - t_1} \quad \text{and} \quad r = \frac{T_1 - T_2}{t_2 - t_1}$$

Factor F is now written, if the cold fluid is circulating in the tubes, as:

$$F = \frac{\sqrt{r^2 + 1}\,\mathrm{Ln}\left[(1-e)/(1-re)\right]}{(r-1)\,\mathrm{Ln}\left[\dfrac{2 - e\left(1 + r - \sqrt{r^2+1}\right)}{2 - e\left(1 + r + \sqrt{r^2+1}\right)}\right]}$$

This is equation 6 in [BOW 40].

This expression can be applied to a large number of passes on the tube side if this number is an even number.

When the exchange is too big (θ big), we can:

1) Put multiple identical machines in series on both liquids.

2) Or choose only one exchanger with multiple passes on shell side, reserving at least two tubes for each of the shell. in any case, the number of tubes per shell must be even, i.e. in pairs.

In the two previous cases, factor F correcting the MLTD calculated on the whole is given by:

$$F = \frac{\sqrt{r^2+1}\,Ln\left[(1-p)/(1-rp)\right]}{(r-1)\,Ln\left[\dfrac{2-p\left(1+r-\sqrt{r^2+1}\right)}{2-p\left(1+r+\sqrt{r^2+1}\right)}\right]}$$

With:

$$p = \frac{1-\left[\dfrac{re-1}{e-1}\right]^{1/N}}{r-\left[\dfrac{re-1}{e-1}\right]^{1/N}}$$

These are equations 8 and 9 in Bowman *et al.* [BOW 40].

N is the number of passes on the shell side, or the number of exchangers in a series. If N = 1, then p = e.

When r = 1, the previous expressions become indeterminate, and we must use:

$$F = \frac{p\left[\sqrt{r^2+1}/(1-p)\right]}{Ln\left[\dfrac{2-p\left(1+r-\sqrt{r^2+1}\right)}{2-p\left(1+r+\sqrt{r^2+1}\right)}\right]}$$

With:

$$p = e / (N - Ne + e)$$

NOTE.–

Exchangers with multiple passes on the tube side are characterized by the fact that crossover may occur (where T = t).

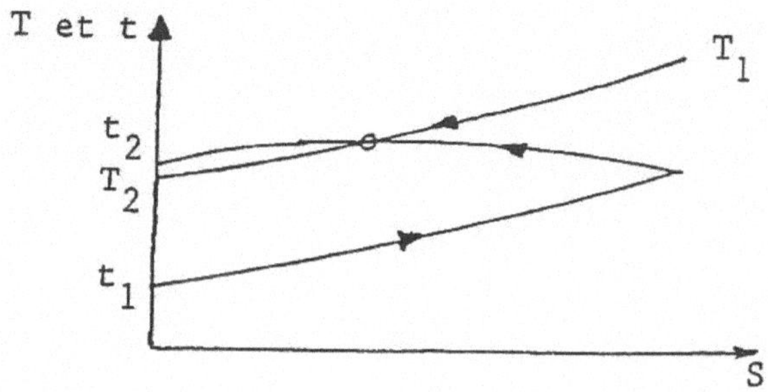

Figure 5.5. *Crossover*

5.2. Fouling

5.2.1. *Heat transfer design*

Some suggestions:

1) In order to preserve the exchanger material, the device may be used at a temperature considerably lower than the input temperature of the hot fluid, by recycling part of the cooled hot fluid exiting the device.

2) An exchanger with a large surface of noble material is expensive. Cheaper materials may be used on the condition that the nature of the operating fluid is also changed. To do this, an intermediate loop must be used, including the new fluid and an additional high-coefficient exchanger (plate exchanger), placed between the fluid in the loop and the initial, corrosive operating fluid.

3) When a product travels through multiple exchangers in a series, savings may be made by rendering the pressure loss in each exchanger proportional to the number of transfer units of the exchanger in question.

5.2.2. *Common types of fouling*

Fouling is the formation of any type of undesirable deposit on a heat transfer surface.

Fouling is more common on rough walls. Evaporator tubes have a rough surface in order to facilitate boiling, as steam bubbles more easily in hollows in the surface. Evaporators are subject to high levels of fouling when used with food products (milk, fruit juices, etc.) and must be cleaned on a daily basis using an appropriate solution (soda, citric acid, detergent, etc.).

When processing suspensions, deposits inevitably occur in dead zones, if these zones are present. Spiral exchangers, which do not have dead zones, are specifically designed for use with liquid suspensions. If a blockage starts to form in a channel, the available cross-section decreases; this causes the speed of the liquid to increase, which then carries off the deposit. Suspensions in air (aerosols) of carbon, cement, pollen or other dust create fouling in air-cooler tubes; consequently, these tubes must be cleaned using pressurized air jets.

The solubility of calcium sulfate increases as the temperature increases; crystallization of the product in hot tubes can be prevented by the addition of polyphosphates, or by demineralizing the water using ion-exchange resins.

Similarly, increased temperature can lead to polymerization and the appearance of tar deposits in the reboiler tubes of certain distillation columns. Kiln tubes in oil refineries are also subject to the formation of coke, reducing the cross section of the tubes, which is even more problematic.

During crystallization of a solution by cooling, the difference in temperature between the liquid and the wall must be limited to avoid crusting, i.e. the formation of a layer of crystals.

If corrosion occurs and results in deposit formation, these deposits may hinder heat transfer; if deposits then become detached from the wall, they may be transported to another location in the device and cause blockages.

Bacteria and algae multiplying within a system can produce a form of sticky slurry, which must be removed using titanium or polypropylene brushes, moved by the circulation of the liquid. Micro-organisms may also be eliminated by regular injections of chlorine into the system.

Hard water is damaging when used as a cooling fluid, as limescale will form on contact with the hot wall if the temperature of this wall exceeds 50°C.

Hard deposits sometimes need to be eliminated manually by cleaning the insides of the tubes with a metal rod after disassembling the exchanger.

For moderate temperatures and low-adherence deposits, fouling may be avoided by pushing foam rubber balls through the water tubes. These balls have a diameter between 1 and 3 mm greater than the internal diameter of the tubes, and are retrieved by a filter on output in the form of a concentrated dispersion, before being reinjected into the system.

Generally, the thickness of deposits increases more or less proportionally over time. However, less consistent deposits may be carried off by the liquid flow, and this constitutes a significant limiting factor. Deposits within a tube limit the cross-section available for the passage of liquid, and so the velocity of this liquid increases; this results in forcible removal (entrainment) of the deposits, stabilizing fouling levels. However, if the fluid is able to flow through multiple tubes, the flow through "clean" tubes will evidently be higher, attenuating this effect. Spiral exchangers, with a single flow channel, present a significant advantage in this respect.

5.2.3. *Practical consequences of fouling*

Heat-transferring walls have a thickness of between 1 and 10 mm, depending on the level of pressure and the nature of the material. The thermal conductivity of the most widespread materials is given in Table 5.1.

Deposits may reach a thickness similar to that of the wall, and, moreover, they have a much lower conductivity. This value varies from 0.5 $W.m^{-1}.°C^{-1}$ (slurries) to 3 $W.m^{-1}.°C^{-1}$, demonstrating the scale of the problem caused by fouling with respect to heat transfer.

Material	λ ($W.m^{-1}.°C^{-1}$)
Mild steel	50
Stainless steel	18
Graphite (with any resin)	90

Table 5.1. *Conductivity of tube materials*

5.2.4. *Typical fouling values*

Typical fouling values are given in Table 5.2.

Fluid	Fouling resistance $10^4 R_f$ (R_f in $m^2.°C.W^{-1}$)
Liquids:	
Fuel oil	10
Transformer oil	2
Lubricating oil	2
Coolant	2
Heat transporter fluids	2
Molten salts	1
Caustic solutions	4
Vegetable oil	6
River water	2–20 (depending on cleanliness)
Well water	1.5
Sea water	1
Gas:	
Exhaust gas	20
Compressed air	4
Organic vapors (including solvents)	2
Acid gases	2

Table 5.2. *Fouling resistances*

5.2.5. *Calculation of wall temperatures*

Fouling generally depends on the temperature of the wall.

This temperature may be obtained by writing the thermal flux density in two ways. For example, for the hot side of the wall:

$$\alpha_T \left(T - t_{pT}\right) = U\left(T - t\right)$$

T and t: temperature of hot and cold fluids (°C)

t_{pT}: temperature of hot wall (C)

$$t_{pT} = T - \frac{U}{\alpha_T}(T - t)$$

Similarly, for the cold side, the wall temperature will be:

$$t_{pt} = t + \frac{U}{\alpha_t}(T - t)$$

6

Condensers, Traps and Condensate Lines

6.1. Surface condensers

6.1.1. *Structure*

Surface condensers are generally tubular devices.

To increase the speed of the coolant, condensers include two passes on the tube side. The tubes have a length twice that of the exchanger, and are bent into a hairpin shape, with coolant input and output points at the same end of the exchanger.

6.1.2. *Assembly*

All distillation columns include a condenser at the top, which receives vapor from the column head. The condensate is then collected in a hotwell.

Condensate is separated into two elements (see Figure 6.1):

– reflux, which is returned to the column;

– distillate, which is extracted.

6.1.3. *Isothermal condensation coefficient of pure vapor*

1) Droplet condensation:

Bodies condense more easily into droplets if they have a high surface tension. Pure water at 20° has a surface tension of 0.072 $N.m^{-1}$; the surface tension of organic bodies is rarely greater than 0.035 $N.m^{-1}$.

This is the reason why only water vapor condenses into droplets on industrial metal surfaces. The coefficients involved in these cases are very high, and in practice, values of between 4,000 and 13,000 $W.m^{-2}.°C^{-1}$ are used. For a vapor of average quality, a value of 7,000 $W.m^{-2}.°C^{-1}$ is advised. Evidently, this coefficient may be reduced to 3,500 $W.m^{-2}.°C^{-1}$ in the case of evaporation vapors (which always include inert elements, which do not condense).

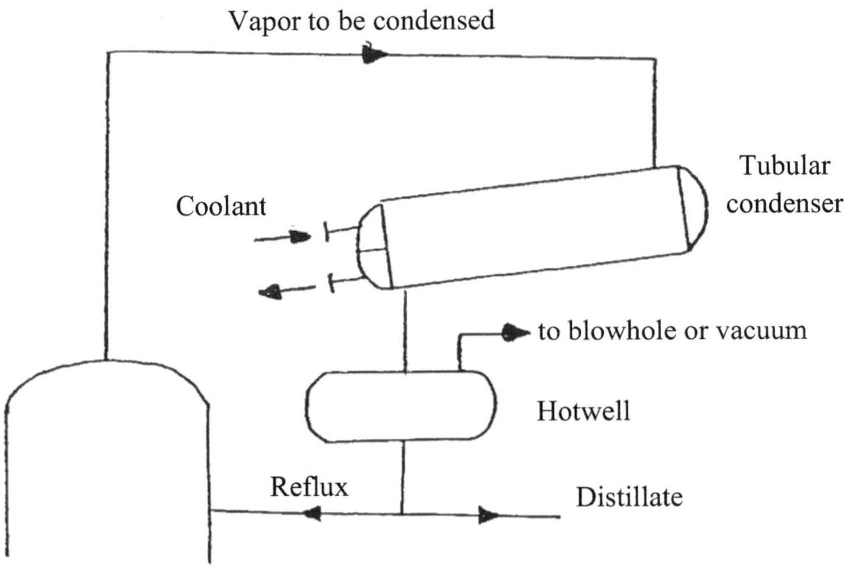

Figure 6.1. *Distillation condenser*

For a metallic surface coated with a product which is not wetted by condensate, condensation will take the form of droplets, but the heat resistance of the coating cancels out the benefits associated with droplet condensation. Moreover, coatings of this type are expensive.

2) Film condensation:

This hypothesis is used for the condensation of any product in the chemical industry, with the exception of water vapor.

For the condensate films let us define the following thickness parameter:

$$\delta = \left[\frac{\mu^2_L}{\rho_L^2 g}\right]^{1/3}$$

μ_L: viscosity of liquid (Pa.s)

ρ_L: bulk density of liquid ($kg.m^{-3}$)

g: acceleration due to gravity (9.81 $m.s^{-2}$).

Using the international system of notation, δ is measured in meters and is of the order of 10^{-4} m.

It is now possible to define the Nusselt number:

$$Nu = \frac{\alpha\delta}{\lambda_L}$$

A Reynolds number:

$$R_e = \frac{4\Gamma}{\mu_L} \text{ with } \Gamma = \frac{W}{P}$$

α: heat transfer coefficient ($W.m^{-2}.°C^{-1}$)

λ_L: thermal conductivity of liquid ($W.m^{-1}.°C^{-1}$)

Γ: load per unit wetted perimeter ($kg.m^{-1}.s^{-1}$)

W: condensate flowrate ($kg.s^{-1}$)

P: perimeter wetted by the condensate (m).

We shall now consider the three main cases encountered in practice:

a) Condensation on a vertical wall:

Duckler [DUC 59] provides graphs for the definition of Nu.

b) Condensation on the outside of a bundle of horizontal tubes:

The corresponding formula may be calculated theoretically. Detailed calculations may be found in Kern [KER 50, p. 261].

The formula is written:

$$\mathrm{Nu} = 1.51\ \mathrm{Re}^{-1/3}; \quad \mathrm{Re} = \frac{4\Gamma}{\mu}$$

The load per unit length is defined, following Kern, by:

$$\Gamma = \frac{W}{N_T^{2/3} L_T}$$

N_T: number of tubes

L_T: tube length (m).

The exponent 2/3, which does not appear in the theoretical formula, takes account of dripping from one tube into another.

c) Condensation inside a bundle of horizontal tubes.

In practice, bundles are placed on a 5–10% slope in order to allow the condensate to drain away.

According to Kern [KER 50], the previous formula remains valid, but the load per unit length is defined as follows:

$$\Gamma = \frac{W}{0.5\ N_T L_T}$$

6.1.4. *Calculation of a pure vapor condenser*

Supposing that the pure vapor for condensation is at a phase change temperature t_b, the heat flow density of condensation is written as:

$$\varphi_c = N\Lambda = U(t_b - t_r)$$

N: flow density of condensed kilomoles ($kmol.m^{-2}.s^{-1}$)

Λ: latent heat of the vapor ($J.kmol^{-1}$)

t_r: coolant temperature (°C).

U is the overall heat transfer coefficient:

$$\frac{1}{U} = \frac{1}{\alpha_c} + \frac{e_p}{\lambda_p} + \frac{1}{\alpha_r}$$

α_c: heat transfer coefficient of condensate along the wall ($W.m^{-2}.C^{-1}$)

α_r: heat transfer coefficient of coolant along the wall ($W.m^{-2}.°C^{-1}$)

e_p: wall thickness (m)

λ_p: thermal conductivity of wall ($W.m^{-1}.°C^{-1}$).

The condensation surface in an elementary section of the device is therefore:

$$\Delta S_c = \Delta Q_c / \phi_c$$

ΔQ_c is the thermal power for condensation in the section.

The temperature rise in the coolant is:

$$\Delta t_r = \Delta Q_c / (W_r C_{pr})$$

W_r: coolant flowrate ($kg.s^{-1}$)

C_{pr}: mass thermal capacity of the coolant ($J.kg^{-1}.°C^{-1}$).

If the vapor is superheated to a temperature $t_G > t_b$, the latent heat flow density is written as:

$$N\Lambda = U(t_b - t_r)$$

U: heat transfer coefficient between the vapor phase and the coolant ($W.m^{-2}.°C^{-1}$).

In an elementary section of the device, the cooling Δt_G of the vapor is written as:

$$W_G C_{pG} \Delta t_G = \Delta S U (t_G - t_b)$$

6.1.5. *Heat flow densities for mixed vapors*

Pure vapors condense at a constant temperature. In all other cases (mixed vapors, presence of non-condensable elements), the temperature varies.

The heat evacuated by the coolant comes from the gaseous phase, which cools and condenses, and from the liquid phase (condensate) which cools down. Let ϕ_G be the heat flow density of the gaseous phase and ϕ_L the flow density of the cooling condensate.

The flow density ϕ_G includes two components:

– a convection term due to contact between gas and the surface of the condensate:

$$\alpha_G (t_G - t_c)$$

t_G: temperature of the gaseous state (°C)

t_c: temperature of the condensate (°C).

– a term corresponding to the flow of molecules for condensation:

$$N\Lambda$$

N: flow density of condensed molecules ($kmol.m^{-2}.s^{-1}$)

Λ: difference in enthalpy between the gas at t_G and the condensate at t_c ($J.kmol^{-1}$).

The heat flow density lost by the gas is:

$$W_G C_G (t_G - t_c) + N\Lambda \quad [6.1]$$

t_c: temperature of condensate (°C).

The same notation is used as in the case of a pure vapor.

6.1.6. *Case of condensable vapor in the presence of non-condensable elements*

Let the vapor be characterized by index A, and the non-condensable elements by an index j varying from 1 to n.

In this case, we use the Maxwell–Stefan equation:

$$C_T \frac{dy_A}{d\delta} = \sum_{j=1}^{n} \frac{y_A N_j - y_j N_A}{D_{Aj}}$$

C_T: overall concentration ($kmol.m^{-3}$)

y_A and y_j: molar fractions of vapor and non-condensable elements in the gaseous phase

N_A and N_j: molar flow densities of vapor A and non-condensable elements ($kmol.m^{-2}.s^{-1}$)

δ: thickness of the boundary layer of gas in contact with the condensate (m)

D_{Aj}: diffusivity of vapor A in the non-condensable j ($m^2.s^{-1}$)

As the non-condensable flows are null, $N_j = 0$.

The Stefan equation is written:

$$C_T \frac{dy_A}{d\delta} = - \sum_{j=1}^{n} \frac{y_j N_A}{D_{Aj}}$$

Take:

$$D_{Am} = \frac{1-y_A}{\sum_{j=1}^{n} \left[\frac{y_j}{D_{Aj}} \right]}$$

If n = 1, we may verify:

$$y_j = 1 - y_A \quad \text{and} \quad D_{Am} = D_{A1}$$

The equation is then written:

$$C_T \frac{dy_A}{d\delta} = -(1-y_A)\frac{N_A}{D_{Am}}$$

Integrating over the thickness δ:

$$N_A = C_T \frac{D_{Am}}{\delta} \text{Ln}\left[\frac{1-y_{Ac}}{1-y_A} \right]$$

y_{Ac}: molar fraction of vapor A in the gaseous phase on the surface of the condensate and in equilibrium with the condensate.

6.1.7. *Mixture of two condensable vapors*

When considering the condensation of two vapors with very different volatilities, the most volatile vapor may be considered to be non-condensable, but this approach is excessively pessimistic. Examples of vapor mixtures with very different volatilities include:

– water and ammonia;

– water and hydrochloric gas;

– methanol and ethylene-glycol, etc.

For vapors A and B, the Maxwell–Stefan equation is written as:

$$C_T \frac{dy_A}{d\delta} = \frac{y_A N_B - (1 - y_A) N_A}{D_{AB}}$$

Take:

$$D_{AB} C_T \frac{dy_A}{d\delta} = (N_A + N_B)\left[y_A - \frac{N_A}{N_A + N_B} \right]$$

And:

$$\frac{N_A}{N_A + N_B} = x_{Ac}$$

x_{AC} is the molar fraction of A in the condensed liquid.

Moreover:

$$N_A + N_B = N$$

Integrating over the thickness δ of the boundary layer:

$$N = \frac{D_{AB}}{\delta} C_T \mathrm{Ln}\left[\frac{x_{Ac} - y_{Ac}}{x_{Ac} - y_A} \right]$$

As before, the heat flux density equation is written as:

$$K(t_c - t_r) = K_G (t_G - t_c) + N_A \Lambda_A + N_B \Lambda_N \qquad [6.2]$$

With:

$$N_A\Lambda_A + N_B\Lambda_B = N\left[x_{Ac}\Lambda_A + \left(1 - x_{Ac}\right)\Lambda_B\right] = N\Lambda$$

6.1.8. *Evaluation of D_{am}/δ and D_{AB}/δ*

Using the Chilton-Colburn analogy:

$$\frac{\beta}{V_m}Sc^{2/3} = \frac{\alpha_G}{C_P\rho_G V_m}Pr^{2/3}$$

β: material transfer coefficient ($m.s^{-1}$)

Sc: Schmidt number (dimensionless)

V_m: mean velocity of gas current parallel to the surface ($m.s^{-1}$)

C_P: mass heat capacity of the gas ($J.kg^{-1}.°C^{-1}$)

ρ_G: bulk density of the gas ($kg.m^{-3}$)

Pr: Prandtl number (dimensionless)

α_G: heat transfer coefficient ($W.m^{-2}.°C^{-1}$).

Hence:

$$\beta = \frac{\alpha_G}{C_p\rho_G}Lu^{2/3}$$

Lu: Luikov number

Coefficient β is simply D_{Am}/δ or D_{AB}/δ.

Thus:

$$C_T\frac{D}{\delta} = \frac{C_T\alpha_G}{\rho_G C_P}Lu^{2/3}$$

C_T is the number of kilomoles contained in 1 m^3 of gas, and ρ_G is the mass in kilograms of 1 m^3 of gas. Thus:

$$\frac{\rho_G}{C_T} = M_G \text{ average molecular mass of the gaseous phase}$$

Hence:

$$C_T \frac{D}{\delta} = \frac{\alpha_G}{C_P M_G} Lu^{2/3}$$

6.1.9. *Evaluation of y_{Ac} and x_{Ac}*

At the temperature of the condensate t_c, and according to the equilibrium coefficients E_A and E_B:

$$y_{Ac} = E_A x_{Ac} y_{Bc} = E_B x_{Bc}$$

$$y_{Ac} + y_{Bc} = 1$$

Thus:

$$E_A x_{Ac} + E_B (1 - x_{Ac}) = 1$$

Hence:

$$x_{Ac} = \frac{1 - E_B}{E_A - E_B} \quad \text{and} \quad y_{Ac} = E_A x_{Ac}$$

We can then calculate N for a given temperature. By successive approximations, we can obtain t_c from the flow density equation [6.2].

NOTE.–

In the case of an azeotrope, after dividing through by $N_A + N_B = N$, the Maxwell–Stefan equation becomes:

$$y_A - x_A = \frac{D_{AB}}{N} \frac{dC_A}{d\delta}$$

in addition, $y_A = x_A$, so $\frac{dC_A}{d\delta} = 0$

There is no temperature gradient as there is no concentration gradient, so the case is treated in the same way as condensation of a pure vapor.

6.1.10. *Coolant flow*

Condensation slows down considerably if the gaseous and liquid phases, taken together, are at a temperature of 5°C above the input temperature of the coolant.

The flow rate W_r of the coolant is therefore calculated based on the fact that the output temperature t_{r2} of the coolant should be 5°C below the dew point of the gaseous mixture. Let Q_c be the thermal power corresponding to condensation between the dew point and the final temperature of the liquid phase:

$$W_r = \frac{Q_c}{C_{pr}\left(t_{r2} - t_{r1}\right)}$$

6.1.11. *Calculation of the surface of a condenser*

The heating program is split into approximately 10 sections, and the flow equation is solved for each section.

With ΔQ_i the thermal power transferred within a section and ϕ_i the heat flow density in the section, the transfer surface corresponding to this section is:

$$\Delta S_i = \Delta Q_i / \phi_i$$

Note that condensation occurs even if the gaseous mixture is superheated (condensation occurs on a glass of iced water even in very hot weather).

The surface of the condenser is then:

$$S = \sum_{i=1}^{n} \Delta S_i$$

The mass condensed in the section is:

$$N_i \Delta S_i M_G$$

M_G: average molar mass of the condensed gaseous phase ($kg.kmol^{-1}$).

Calculations for each section are carried out in two steps:

– step one, using section input data;

– step two, using data from the middle of the section (arithmetic mean of input and output data).

6.2. Overcooling of the condensate

6.2.1. *General points*

The condensate is at an intermediate temperature between that of the coolant and that of the gaseous phase, but it is possible to bring the condensate down to a temperature closer to that of the coolant.

For this, the coolant is introduced into the lower section of the device, i.e. the section where the condensate accumulates; the lower part of the transfer surface is thus covered with stagnant condensate. A non-return flow device is installed for this purpose at a pre-determined height.

The heat exchange involved in the overcooling process occurs by natural convection, and the transfer coefficient is low.

6.2.2. *Transfer coefficient for overcooling*

When the overcooling condensate is in contact with a vertical surface, the natural convection coefficient may be calculated using the following expression:

$$\alpha = Nu\lambda_L / L_{SR}$$

Nu: Nusselt number (dimensionless)

λ_L: thermal conductivity of the liquid condensate ($W.m^{-1}.°C^{-1}$)

L_{SR}: vertical height of the cooling surface covered by the liquid (m)

$$Nu = 0.5(Pr \times Gr)^{1/4}$$

Pr: Prandtl number:

$$Pr = \frac{C_p \mu_L}{\lambda_L}$$

μ_L: dynamic viscosity of the liquid condensate (Pa.s)

C_p: thermal capacity of the condensate ($J.kg^{-1}.°C^{-1}$)

Gr: Grashof number

$$Gr = \frac{\beta g \Delta T \rho_L^2 L_{SR}^3}{\mu_L^2}$$

g: acceleration due to gravity (9.81 $m.s^{-2}$)

ΔT: temperature difference between the wall and the center of the liquid mass (°C)

β: cubic dilation coefficient of the liquid ($°C^{-1}$)

$$\beta = -\frac{1}{\rho_L}\frac{d\rho_L}{dT}$$

ρ_L: density of the liquid ($kg.m^{-3}$)

$d\rho_L/dT$ may be obtained from the slope of the tangent to the curve representing variations in density as a function of temperature.

EXAMPLE 6.1.–

$\rho_L = 1{,}180\ kg.m^{-3}$ $\qquad$ $Pr = 5$

$\beta = 17 \times 10^{-4} °C^{-1}$ $\qquad$ $g = 9.81\ m.s^{-2}$

$\Delta T = 12°C$ $L_{SR} = 0.5$ m

$\mu_L = 0.45$ Pa.s $\lambda_L = 0.12$ W.m^{-1}.°C^{-1}

$$Gr = \frac{17\times10^{-4}\times9.81\times12\times1180^2\times0.5^3}{0.45^2}$$

$$Gr = 17.5\times10^{10}$$

$$Nu = 0.5\left(5\times17.5\times10^{10}\right)^{1/4}$$

$$Nu = 482$$

$$\alpha = \frac{482\times0.12}{0.5}$$

$$\alpha = 115.7 \text{ W.m}^{-2}.°C^{-1}$$

6.3. Contact condensers

6.3.1. *General points*

A contact condenser is a vertical cylindrical condenser with a height around two to four times its diameter.

Contact condensers are used in the following cases:

– where there is a significant risk of corrosion. Using this method, the wall can be lined with no negative effects on heat transfer;

– where fouling may cause significant problems. Systems of this type are not subject to blockages;

– when the input mixture contains a high proportion of non-condensable elements;

– where the amount of effective heat to evacuate is high;

– when condensing evaporation installation vapors.

However, contact condensers should not be used with foaming products.

The vapor for condensation is introduced via a lateral tube at the base of the device at a velocity of between 15 and 35 $m.s^{-1}$, depending on the pressure of the vapor. The non-condensable components move to the upper section of the device.

In order to control the contact between the vapor and the liquid, the liquid may be sprayed and/or cascaded onto the vapor.

The droplets of liquid on which the vapor has condensed come together at the device output point, and the reconstituted liquid is evacuated through a large tube, connected to the condenser shell by a truncated cone.

The water used to collect condensates from vapors in the upper part of evaporation devices is, in most cases, drained away.

However, different assemblies may be used, particularly in the food industry.

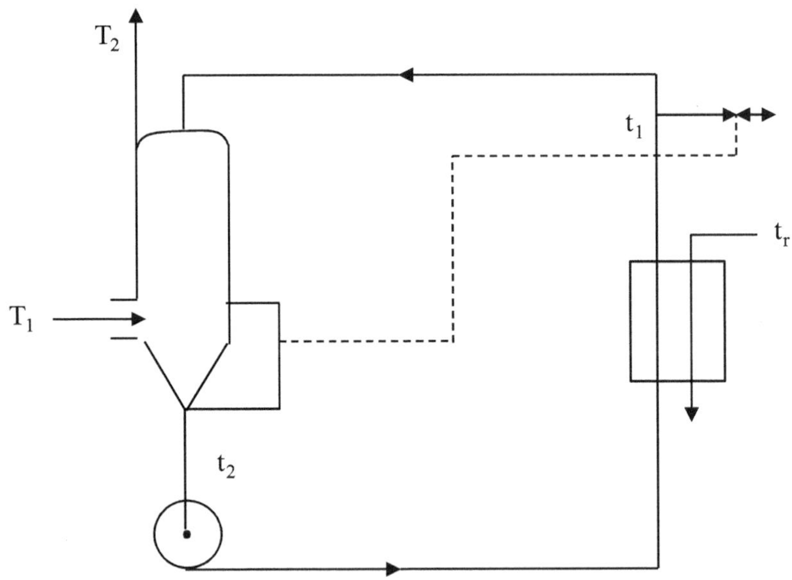

Figure 6.2. *Contact condenser assembly used in the food industry*

The liquid collected in the base of the condenser is retrieved by a pump, which transmits the liquid to an exchanger (generally a plates exchanger), where it is cooled before being returned to the upper section of the condenser.

As the vapor condenses, the volume of liquid in the loop increases, and part of this liquid then needs to be removed.

Condensates are extracted after the exchanger, i.e. at the point where they are coldest.

The output valve is a regulation valve controlled by the level of liquid in the base of the condenser.

6.3.2. *Calculation of contact condensers*

1) Determination of operating temperatures:

The input temperature T_1 of the gaseous phase is imposed, along with the temperature of the cooled liquid, t_1.

The output temperature T_2 of the non-condensable elements is 2–3 degrees higher than temperature t_1. Note that, for a surface condenser, we would have:

$$T_2 - t_1 \geq 4°C$$

The output temperature of the water-condensate mixture, t_2, should be equal to T_1 minus approximately 3°C.

In most practical applications, the difference $t_2 - t_1$ is somewhere between 5 and 15°C, although, naturally, there are exceptions to this rule.

Given that contact condensers allow the use of smaller differences $(T_2 - t_1)$ than surface condensers, they use less water.

In the case of recirculation in an external plate exchanger, we may take:

$$t_1 = t_r + 2$$

t_r is the temperature of the coolant.

2) Heat balance:

The heat lost by the gaseous phase is:

$$Q = W_v (I_v - C_{PL} t_2) + W_I C_{pI} \cdot (T_1 - T_2)$$

With:

W_v: condensed vapor flow (kg.s^{-1})

I_v: enthalpy of vapor (J.kg^{-1})

W_I: non-condensed flow (kg.s^{-1})

C_{pI}: mass heat capacity of non-condensed elements (J.kg^{-1}.°C^{-1}).

The heat gained by the liquid is:

$$Q = W_L C_{pL} \cdot (t_2 - t_1)$$

With:

W_L: liquid flow (kg.s^{-1})

C_{pL}: heat capacity of the liquid (J.kg^{-1}.°C^{-1}).

Taking the two exchanged heat values to be equal, we obtain the value of the input water flow W_L.

3) Condenser geometry:

This calculation includes the following elements:

– exchange surface required:

$$S = \frac{Q}{\alpha \Delta T}$$

Q: exchanged thermal power (calculated in the heat balance) (Watts)

S: exchange surface (m^2)

ΔT: mean logarithmic temperature difference (°C)

α: heat transfer coefficient ($W.m^{-2}.°C^{-1}$).

If the condensed vapor is water vapor, we may take:

$$\alpha = 7000 \exp(-4.94\ y_I)$$

y_I: molar fraction of non-condensable elements in the input vapor phase.

For organic vapors, the coefficient will be three times lower.

The volumetric area a_g of the droplets depends on their diameter. For a contact condenser, we may consider that:

$$d_g = 5 \times 10^{-4}\ m$$

Hence:

$$a_g = \frac{6}{d_g} = 12000\ m^2.m^{-3} = 12000\ m^{-1}$$

The volume of liquid required for the condenser to have an exchange surface S can then be calculated.

$$V_L = S / a_g$$

V_L: volume of liquid present (m^3)

S: exchange surface (m^2)

a_g: volumetric surface of droplets (m^{-1}).

1) Liquid residence time and volume of the condenser

$$\tau = V_L / Q_L$$

τ: residence time (s)

Q_L: liquid flow ($m^3.s^{-1}$).

The residence time of the gas will be considered to be equal to that of the liquid. The volume V of the condenser can now be calculated:

$$V = Q_G \tau$$

Q_G: gaseous phase flow ($m^3.s^{-1}$)

V: volume (m^3)

The dimensions of the condenser are obtained as follows:

The velocity of the liquid in an empty container must be greater than 0.03 $m.s^{-1}$ and less than 0.1 $m.s^{-1}$.

Given a liquid velocity, we can determine the cross-section of the device. Knowing the volume, we can calculate the height, which should be 2–4 times the diameter.

Moreover, the empty container velocity of the gaseous phase input must be less than 10 $m.s^{-1}$. This velocity is calculated over the cross-section of the condenser.

2) Diameter of the barometric column:

The height of this column is 10.33 m. Its diameter must be large enough to avoid the creation of gas pockets by the liquid, which would result in vibrations.

According to Hills the Froude number must be less than 0.3:

$$\mathrm{Fr} = \frac{V}{\sqrt{gD}} < 0.3$$

V: velocity of liquid in tubes ($m.s^{-1}$)

g: acceleration due to gravity (9.81 $m.s^{-1}$)

D: diameter of tubes (m).

This gives a lower limit for D:

$$D > 1.13Q_L^{0.4}$$

Q_L: flow of liquid ($m^3.s^{-1}$)

A safety margin of 20% is added to D, so:

$$D > 1.35Q_L^{0.4}$$

The volume of the hotwell will be defined as twice that of the barometric column.

6.4. Trap types

6.4.1. *Use of condensate traps*

Water vapor loses its latent heat when condensing, and the resulting liquid must be removed so that the heating surface does not become covered. The device used in this case must allow the condensate to escape, but not the vapor. These devices are known as traps.

6.4.2. *Thermostatic bimetallic trap*

A bimetallic strip, as the name suggests, is made up of two strips of different metals which are welded. The difference in thermal dilation between the two metals causes the strip to bend, and this acts, via a rod, on the stopper of the condensate exhaust opening.

This type of trap is not hugely sophisticated, but robust and cheap, and able to withstand both freezing and water hammer. However, it has high thermal inertia and requires overcooling of the condensate, sometimes to a considerable extent (≥ 30°C). The response time of these traps is long, and the opening frequency is low, meaning that the devices are unable to cope with rapid regime changes.

Certain manufacturers produce balancing systems to compensate for the effects of pressure on both sides of the bimetallic strip. These strips have a limited power, and also require regular calibration.

6.4.3. *Balanced liquid expansion thermostatic trap*

A liquid is contained in a capsule with an elongated bellows shape, the length of which varies as the volume of the liquid changes due to changes in temperature. The vent acts on the exhaust shutter via a pin.

The trap possesses significant power, and there is no need for an additional balancing element, as in the case of bimetallic strip traps. Traps of this type are therefore said to be “balanced”.

Liquid expansion traps are frost resistant, and are used on the output side of coils and tracer line.

Periodic calibration is possible using an adjustable overload spring.

6.4.4. *Balanced pressure steam trap*

A liquid contained in a capsule (bellows-shaped or flat) vaporizes when heated, changing the shape of the capsule, which then acts via a rod on the condensate exhaust shutter, as before.

The valve opens if the system is cold, and therefore only *cold* condensates or non-condensable elements are drained. Condensates need to be over-cooled by around 20°C. The valve closes when the temperature of the condensate is around 2°C higher than the valve opening temperature.

The opening temperature can be changed by selecting a liquid with a different boiling temperature.

As with all thermostatic traps, these devices operate in a discontinuous manner, but certain designs are able to maintain a continuous minimum output.

Steam traps are resistant to freezing and can withstand pressures up to 15 bar and temperatures up to 300°C. They are small, and able to evacuate any gas-liquid or non-condensable gas mixture, as long as it is sufficiently cool.

6.4.5. *Inverted bucket steam traps*

An inverted bucket element is placed in a capsule. Liquid arrives in the lower section of the container, under the bucket. If gas also enters the system, the bucket, along with the liquid upon which it floats, will constitute a hydraulic joint for the gas.

Unlike free-float traps, the inverted bucket prevents evacuation of condensate from the container when in its highest position, and allows evacuation when in a low position. Notably, if non-condensable gas arrives in the system, it fills the bucket, which then rises to prevent evacuation. However, a small hole is left in the top of the inverted bucket, allowing the gas (which supports the weight of the bucket) to escape. The bucket then falls, allowing the condensate to be evacuated.

If a condensable vapor arrives in the system, this gas will condense inside the bucket, causing it to descend, opening the condensate evacuation valve. Traps of this type are therefore only able to function if there is sufficient heat loss to cool the condensate present in the container.

If the trap is overdimensioned, or if the pressure of the condensates is too low, then the inverted bucket will remain in the low position and vapor will leak from the device. Traps of this type should therefore be reserved for use with vapors at medium to high pressures.

Inverted bucket traps are resistant to water hammer, but sensitive to freezing. Their air evacuation capacity is limited, and it is advisable to use a thermostatic vent alongside the trap.

Sliding needles are used to prevent fouling, ensuring that the hole at the top of the bucket does not become blocked. The device is only able to function if there is water in the container, meaning that the trap needs to be placed *under* the purging bottle.

Unlike free-float traps, inverted bucket traps operate at a frequency which is independent of events in the piping. Correct operation can be verified using dry saturated vapor, in which case opening should occur every 20–30 s, even if very little condensate is produced. Less than 10°C of overcooling is required.

6.4.6. *Free-float traps*

Containers are fitted with an internal float. As the level of condensate rises, the float also rises, and acts on a lever to pull the valve plug, allowing the evacuation of liquid condensate.

In this case, valve opening is not dependent on pressure or temperature, but is triggered by the level of the liquid. It is therefore not possible to evacuate non-condensed elements, even if these elements are cold. Containers using traps of this type must be fitted with a thermostatic mechanism to evacuate cold gases from the upper section of the cylinder.

Free-float traps offer a rapid response to regime changes. However, condensates will be evacuated even if they are not fully cooled, and the heat of these condensates is lost.

As water is always present in the container, these devices need to be protected from freezing.

Free-float traps are able to cope with high condensate flows, and are not sensitive to back pressure. *They are the only trap type able to operate in a vacuum*, and to evacuate hot liquids or condensates.

6.4.7. *Thermodynamic traps*

These devices make use of the expansion of a liquid (the condensate) with partial vaporization, which justifies the name of "thermodynamic traps". Naturally, if the condensate is sufficiently over-cooled, it will not vaporize.

1) Check valve traps:

Before entering the main body of the trap, the condensate passes through a converging element, where its velocity is increased, to the detriment of the pressure. A mobile disc element is placed above the input and output points, acting as a check valve.

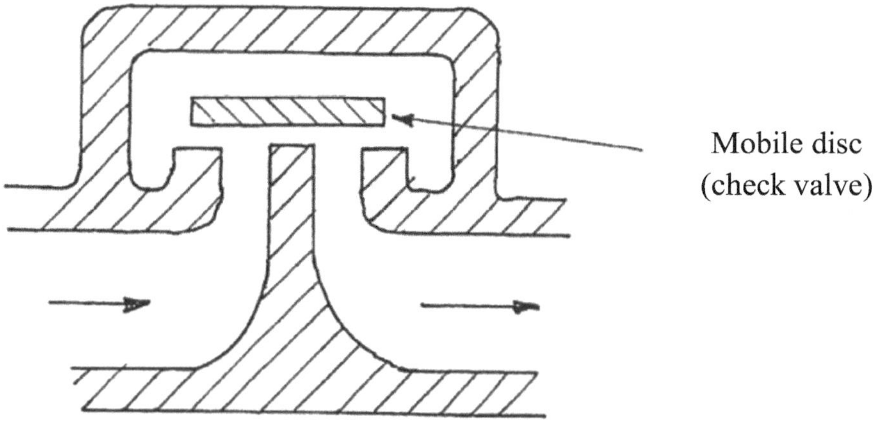

Figure 6.3. *Thermodynamic mobile disc trap*

If the condensate is hot, it vaporizes, and the body of the trap is filled with a pocket of vapor. The disc, which does not float in vapor, descends and closes the condensate inlet and outlet. If cold condensate arrives in the device, the vapor condenses, and the cavity (body of the trap) becomes filled with liquid; the kinetic energy of the incoming liquid is sufficient to lift the disc, opening a passage to the outlet point.

These devices are robust, small in size, able to withstand high pressures, not subject to freezing and self-draining. There is therefore no need for a bypass mechanism.

2) Continuous multi-nozzle "trap":

Multiple nozzles, of increasing diameter, are placed in series, and a screw mechanism with a command wheel is installed in order to adjust the cross-section through which the liquid is able to flow.

"Traps" of this type are used for very high flow rates (several tonnes of vapor per hour) in urban heating systems. The devices cannot be closed, and there is a continuous output of condensates and/or vapor, meaning that they cannot truly be considered to be traps.

6.5. Trap selection and assembly

6.5.1. *Trap installation (classic assembly)*

The drained substance is collected in bottles which empty via traps to avoid vapor loss. Expansion and partial vaporization of the condensate only occurs at the trap outlet if the condensate has not been over-cooled.

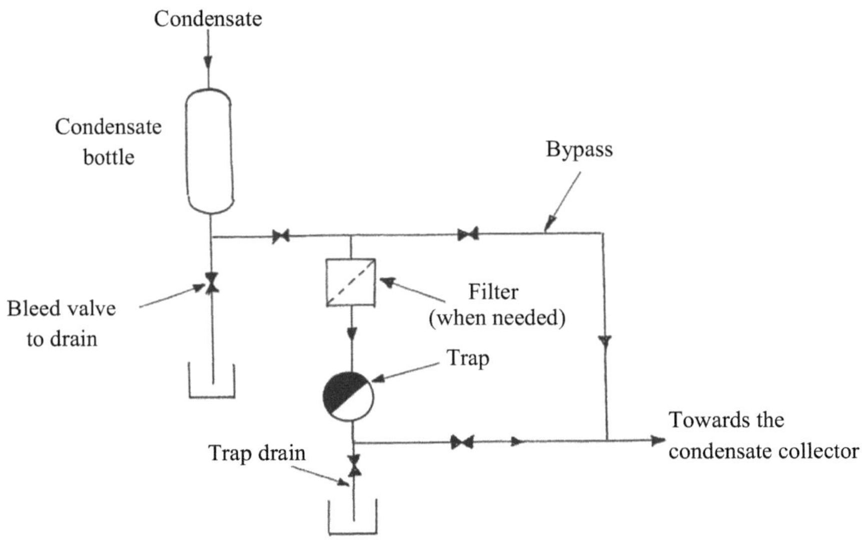

Figure 6.4. *Classic assembly of a condensate trap*

The bottle-trap assembly will not be able to evacuate the desired condensate flow (with consideration given to transitory regimes) if:

– the volume of the bottle (corresponding to the accumulation of condensate between two drainings) is not sufficient;

– the draining frequency (resulting from cooling of the condensate) is not sufficient.

In this case, the condensate will overflow into the piping, and banging will occur due to the sudden condensation of the vapor on contact with the overcooled condensate.

One means of coping with transitory regimes is to add a free-float trap to the thermostatic trap. The thermostatic trap will thus evacuate the permanent flow, whilst the free-float trap will enable the system to cope with sudden peaks in flow.

6.5.2. *Direct draining to the atmosphere*

If the trap does not have sufficient condensate flowrate capacity, it will be unable to cope with peaks in the flow of condensate (for example, on startup).

In this case, free air bleeding is essential, running parallel to the operation of the trap. The bleeding is generally operated manually.

6.5.3. *Bypass to the condensate network*

The use of a bypass to the condensate evacuation network (rather than to the atmosphere) is only necessary if the average condensate flow is particularly high (above 200 $kg.h^{-1}$, for example).

NOTE (Bleeding).–

Piping and devices will be filled with air after a prolonged shutdown, and will need to be bled manually (DN15) on startup. The drainage valve may be left open for around 10 min, until a jet of vapor is observed.

6.5.4. *Banging in vapor pipes*

The presence of over-cooled condensates in a vapor collector causes sudden and noisy condensation of part of the vapor, with the same sound spectrum as a water hammer wave.

If draining does not take place in the inferior bend preceding a vertical ascent, condensate resulting from heat loss can descend the vertical section and accumulate, creating a strong probability of banging in the pipes.

6.5.5. *Flowrate of condensate for drainage in steam pipes*

The permanent flow corresponds to heat loss to the surrounding atmosphere.

The external surface of a pipe with a diameter ϕ (meters) with 5 cm of insulation is equal to π (ϕ + 2 × 0.05) per meter of the length. If the loss coefficient for a pipe of this type is estimated at a maximum of 1 W.m^{-2}.°C^{-1} and T_V is the temperature of the steam, the thermal power loss will be:

$$Q_p = \pi(\phi + 2\times 0.05)(T_V - 20)\left(\text{Watt.m}^{-1}\right)$$

As the latent heat of steam condensation L_c is of the order of 2×10^6 J.kg^{-1}, the permanent flow of condensate produced will be:

$$W_p = Q_p / L_c \left(\text{kg.s}^{-1}.\text{m}^{-1}\right)$$

The transitory additional condensate flow is the result of the pipe heating as the temperature (and therefore pressure) of the steam increases.

The temperature of the steam will thus increase by δT_V over the time interval $\delta\tau$. Furthermore, let M_L be the mass of the pipe per meter of length, and 460 J.kg^{-1}.°C^{-1} the heat capacity of steel. The corresponding transitory power will be:

$$Q_T = \frac{460\ M_L \delta T_V}{\delta\tau}\left(\text{Watt.m}^{-1}\right)$$

The value of M_L may be estimated using the table below:

DN (mm)	M_L (kg.m^{-1})
200	30
300	55
500	165

As the latent heat of vapor condensation L_c is of the order of 2×10^6 J.kg^{-1}, the total flow of condensate for draining per meter of pipe is:

$$W_C = (Q_P + Q_T)/L_c \left(kg.s^{-1}.m^{-1}\right)$$

EXAMPLE 6.2.–

$\phi = 0.2$ m $\quad T_V = 180°C \quad L_c = 2\times 10^6$ J.kg^{-1}

$\delta T_V = 20°C \quad \delta\tau = 15$ mn $= 900$ s $\quad M_L = 30$ kg.m^{-1}

$$Q_p + Q_T = \pi(0.2+0.1)160 + 460\times 30\times 20/900 = 150.8 + 316.9$$

$$W_C = 2.34\times 10^{-4} kg.s^{-1}.m^{-1} = 0.84\ kg.h^{-1}.m^{-1}$$

6.5.6. *Trap selection*

In practice, the choice of trap design is imposed by the steam pressure P_V and the condensate flowrate W_C.

1) $P_V > 2$ bar G and $W_C < 150$ kg.h^{-1}

A thermodynamic trap is recommended (small vapor collectors, tracing lines, certain double-wall devices, pilot exchangers).

2) $P_V > 2$ bar G and 150 kg.h$^{-1} < W_C < 5{,}000$ kg.h^{-1}

Inverted bucket traps or, in the absence of rapid fluctuations, thermostatic traps may be used for processing equipment within this range.

3) Any P_V and $W_C > 5{,}000$ kg.h^{-1}

In this case, a free-float trap or even a condensate hotwell (filled by gravity) with level regulation may be used.

4) $P_V < 2$ bar G and $W_C < 5{,}000$ kg.h^{-1}

An inverted bucket (a "bell") trap or a thermostatic device may be used, may be as a free-float trap, depending on the possibility of over-cooling. The sum of heat losses contributing to cooling the condensate (in Watt) is:

$$Q_{RC} = 150H + 70(L_1 + L_2) + 120 + (\text{potentially})230$$

H: height difference between the lower generator of the vapor tube and the drain valve (m)

L_1: length of connection between the drain valve and the bottle (m)

L_2: length of connection between the trap and the drain valve (m).

The value of 230 W corresponds to the heat loss from a bottle, if a bottle is used.

Taking W_C as the mass flowrate of condensates processed by the trap, the over-cooling in relation to the saturation is, in degrees Celsius:

$$\delta t = Q_{RC}/4180\ W_C \left(W_C \text{ in kg.s}^{-1}\right)$$

(4,180 is the heat capacity of water in $J.kg^{-1}.°C^{-1}$).

$\delta t < 2°C$: a free-float trap is required.

$2°C < \delta t < 10°C$: an inverted bucket trap may be used.

$30°C > \delta t \geq 10°C$: a thermostatic trap may be used.

6.6. Condensate lines

6.6.1. *Evacuation and use of condensates*

Condensate collection pipes are often designed to operate using gravity, as the trap output pressure is similar to that of the atmosphere.

Moreover, the pressure loss through friction must be avoided in order to prevent the condensates from expanding, forming vapor pockets which impede drainage.

A velocity of 0.2 $m.s^{-1}$ is used as a maximum velocity when calculating condensate piping designs.

Instead of traveling through a trap, vapor condensates may also be expanded in order to produce low-pressure vapor and hot water (for example, steam at 1.2 bar G and water at 105°C). The fraction x of the condensate which is vaporized is such that:

$$h_C = xH_D + (1-x)h_D$$

h_C: mass enthalpy of condensates (kJ.kg^{-1})

H_D and h_D: mass enthalpy of expanding vapor and hot water obtained in this way (kJ.kg^{-1})

The enthalpy values H_D and h_D are calculated at saturation point, and again at the expanded pressure.

This expansion takes place in capacities (with an elongated vertical shape and a rounded base).

Condensates are often considered to be corrosive, as they are often produced from steam containing no NH_3 gas or cyclohexylamine vapor; in the presence of even a small amount of CO_2, these condensates become acidic and corrosive.

7

Boiling and Heat Transfer

7.1. Stagnant boiling

7.1.1. *Definitions*

Stagnant boiling concerns the vaporization of a liquid on contact with a heated wall, *without* the liquid, as a whole, moving in relation to the wall.

Convective boiling, on the other hand, pertains to a liquid which boils whilst moving along a heated wall.

For a moderate variation ΔT in the temperature of a liquid, the saturation pressure of the liquid varies by $\Delta\pi$.

$$\Delta\pi = \frac{L\rho_v}{T_L}\Delta T = H\Delta T \qquad [7.1]$$

L: latent vaporization heat ($J.kg^{-1}$)

ρ_v: density of vapor ($kg.m^{-3}$)

T_L: temperature of liquid mass (K)

$$H = L\rho_v / T_L$$

$$\Delta T = T_p - T_L$$

T_p: wall temperature (K)

Equation [7.1] is the *Clapeyron formula.*

7.1.2. *Density of heat flowrate through the wall*

Forster and Zuber's equation [FOR 55], modified by Greif, gives an expression of the heat flow density ϕ_T, and we can write:

$$\phi_T = K\Delta T^2$$

With:

$$K = 0.000734\,\rho_v L\left[\frac{Ha}{2\sigma}\right]^{0.5}\times\left[\frac{H}{\rho_L}\right]^{0.25}\times\left[\frac{12\rho_L a}{\pi\mu_L}\right]^{0.625}\times\left[\frac{\rho_L C_L}{\rho_v L}\right]^{1.25}\times Pr^{1/3}$$

ρ_L and ρ_v: densities of liquid and vapor (kg.m^{-3})

L: latent vaporization heat (J.kg^{-1})

a: thermal diffusivity of liquid (m^2.s^{-1})

$$a = \frac{\lambda_L}{C_L\rho_L}$$

π: 3.1416

σ: surface tension of liquid (N.m^{-1})

λ_L: thermal conductivity of liquid (W.m^{-1}.°C^{-1})

C_L: thermal capacity of liquid (J.kg^{-1}.°C^{-1})

μ_L: viscosity of liquid (Pa.s)

Pr: Prandtl number of liquid

$$Pr : \frac{C_L\mu_L}{\lambda_L}$$

$$H = \frac{L\rho_v}{T_L}$$

Quantities L, μ_L and ρ_v are evaluated at T_L.

$$\Delta T : T_p - T_L$$

T_p: absolute temperature of the wall (K)

T_L: absolute temperature of liquid mass (K).

EXAMPLE 7.1.–

$C_L = 4{,}180\ \text{J.kg}^{-1}.°\text{C}^{-1}$ $\sigma = 0.072\ \text{kg.s}^{-2}$

$\rho_L = 1{,}000\ \text{kg.m}^{-3}$ $T_L = 100\ \text{C} = 373\ \text{K}$

$\mu_L = 0.25\times 10^{-3}\text{Pa.s}$ $L = 2.26\times 10^{6}\ \text{J.kg}^{-1}$

$\lambda_L = 0.64\ \text{W.m}^{-1}.°\text{C}^{-1}$ $\rho_v = 0.6\ \text{kg.m}^{-3}$

$$a = \frac{0.64}{4180\times 1000} = 1.53\times 10^{-7}\,\text{m}^2.\text{s}^{-1}$$

$$H = \frac{2.26\times 10^{6}\times 0.6}{373} = 3635.4\ \text{Pa.°C}^{-1}$$

$$K = 0.000734\times 0.6\times 2.26\times 10^{6}\times\left[\frac{3635.4\times 1.53\times 10^{-7}}{2\times 0,072}\right]^{0.5}\times\left[\frac{3635.4}{1000}\right]^{0.25}$$

$$\times\left[\frac{12\times 1000\times 1.53\times 10^{-7}}{0.25\times 10^{-3}\times\pi}\right]^{0.625}\times\left[\frac{1000\times 4180}{2.26\times 10^{6}\times 0.6}\right]^{1.25}\times\left[\frac{4180\times 0.25\times 10^{-3}}{0.64}\right]^{0.33}$$

$$K = 699\ \text{W.m}^{-2}.°\text{C}^{-2}$$

7.1.3. *Calculation of overheating of the wall*

We know that the heat flowrate density ϕ_T is:

$$\phi_T = K\Delta T^2$$

From this, the partial transfer coefficient may be deduced:

$$\alpha_{eb} = \frac{\phi_T}{\Delta T} = K\Delta T$$

Giving the overall coefficient, U:

$$\frac{1}{U} = \frac{1}{\alpha_S} + \frac{e_p}{\lambda_p} + R_e + \frac{1}{\alpha_{eb}}$$

α_S: coefficient on the operating fluid side (generally water vapor) ($W.m^{-2}.°C^{-1}$)

e_p: wall thickness (m)

λ_p: thermal conductivity of the wall ($W.m^{-1}.°C^{-1}$)

R_e: fouling resistance ($m^2.°C.W^{-1}$)

α_{eb}: partial coefficient on the boiling side ($W.m^{-2}.°C^{-1}$)

$$\frac{1}{U} = A + \frac{1}{K\Delta T} \quad \text{with}: A = \frac{1}{\alpha_S} + \frac{e_p}{\lambda_p} + R_e$$

And:

$$\frac{1}{\phi_T} = \frac{1}{U(T_S - T_L)} = \frac{1}{(T_S - T_L)}\left(A + \frac{1}{K\Delta T}\right) = \frac{1}{K\Delta T^2}$$

The last equality is written as:

$$AK\Delta T^2 + \Delta T - (T_S - T_L) = 0$$

i.e. taking the positive square root of this equation:

$$\Delta T = \frac{-1 + \sqrt{1 + 4(T_S - T_L)AK}}{2AK}$$

T_S: temperature of heating fluid (K)

T_L: temperature of liquid (K).

Finally:

The flow density for stagnant boiling is:

$$\phi_T = K\,\Delta T^2$$

The corresponding heat transfer coefficient is:

$$\alpha_{eb} = \frac{\phi_T}{\Delta T} = K\,\Delta T$$

With:

$$\Delta T = T_p - T_L$$

EXAMPLE 7.2.–

$K = 699\ W.m^{-2}.°C^{-2}$ $\quad R_e = 0$

$\alpha_S = 7{,}500\ W.m^{-2}.°C^{-1}$ $\quad T_S = 120\ °c = 393\ K$

$e_p = 0.002\ m$ $\quad \lambda_p = 15\ W.m^{-1}.°C^{-1}$

$T_L = 373\ K$

$$A = \frac{1}{7500} + \frac{0.002}{15} = 2.66\times10^{-4}\,m^2.°C.W^{-1}$$

$$\Delta T = \frac{-1+\sqrt{1+4\times20\times699\times2.66\times10^{-4}}}{2\times699\times2.66\times10^{-4}} = 8.02\,°C$$

$$\phi_T = 699\times8.02^2 = 45018\ W.m^{-2}$$

$$\alpha_{eb} = 699\times8.02 = 5605\ W.m^{-2}.°C^{-1}$$

7.1.4. *Drying of the wall*

As the temperature of the operating liquid is increased, the temperature T_p of the wall also increases. This results in an increase in ΔT, since:

$$\Delta T = T_p - T_L$$

T_L: temperature of the liquid mass.

The density of the transmitted heat flow ϕ_T as a function of ΔT passes via a maximum.

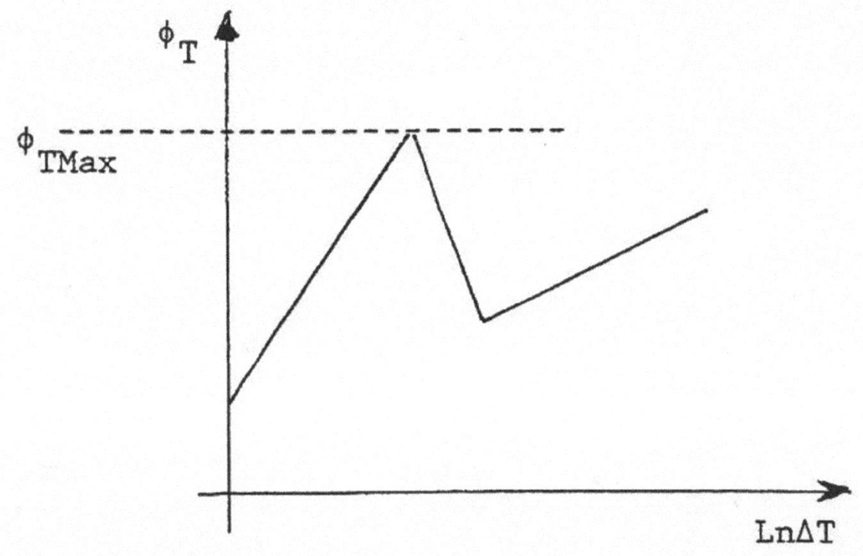

Figure 7.1. *Heat flow density as a function of ΔT*

The value ϕ_{TMax} is due to the creation of bubbles, which limit the contact of the liquid with the wall and thus limit the transmitted heat. This leads to calefaction and the wall is, locally, dried out.

The expression of the maximum flow can be obtained using a mnemotechnical method.

The Newton number is obtained by:

$$\mathrm{Ne} = \frac{\text{force of gravity}}{\text{resistance of fluid}} = \frac{(\rho_L - \rho_v)\ell^3 g}{\ell^2 \rho_v V^2} = \frac{\ell(\rho_L - \rho_v)g}{\rho_v V^2}$$

ℓ: characteristic dimension (m)

The Weber number is:

$$\mathrm{We} = \frac{\text{resistance of fluid}}{\text{capillary force}} = \frac{\ell^2 \rho_v V^2}{\sigma \ell} = \frac{\ell \rho_v V^2}{\sigma}$$

The maximum flow should not be dependent on the characteristic length ℓ. Let us eliminate this length:

$$\frac{\mathrm{We}}{\mathrm{Ne}} = \frac{\rho^2_v V^4}{\sigma(\rho_L - \rho_v)g} \quad \text{hence}: V = \left(\frac{\sigma(\rho_L - \rho_v)g}{\rho_v^2}\right)^{1/4} \times \left(\frac{\mathrm{We}}{\mathrm{Ne}}\right)^{1/4}$$

And:

$$\phi_{\mathrm{TMax}} = L\rho_v V = \left(\frac{\mathrm{We}}{\mathrm{Ne}}\right)^{1/4} \times L\, \rho_v^{1/2} \times \left(\sigma(\rho_L - \rho_v)g\right)^{1/4}$$

Based on Kutateladse's tests [KUT 51]:

$$\left(\frac{\mathrm{We}}{\mathrm{Ne}}\right)^{1/4} = 0.13 \# \frac{\pi}{24}$$

Finally:

$$\phi_{\mathrm{TMax}} = 0.13\; L\rho_v^{1/2}\left[g\sigma(\rho_L - \rho_v)\right]^{1/4}$$

L is the latent vaporization heat ($\mathrm{J.kg^{-1}}$).

EXAMPLE 7.3.–

In the case of water:

$L = 2.26\times 10^6\ J.kg^{-1}$ $g = 9.81\ m.s^{-2}$

$\rho_v = 0.6\ kg.m^{-3}$ $\sigma = 0.072\ kg.s^{-2}$

$\rho_L = 1{,}000\ kg.m^{-3}$

$$\phi_{TMax} = 0.13\times 2.26\times 10^6 \times 0.6^{1/2}\left(9.81\times 0.072\times 1000\right)^{1/4}$$

$$\phi_{TMax} = 1170641\ W.m^{-2}$$

7.2. Convective boiling

7.2.1. *Heat transfer coefficient*

Let us calculate the parameter of the physical properties:

$$\psi = \left[\frac{\rho_G}{\rho_L}\right]^{0.5} \times \left[\frac{\mu_L}{\mu_G}\right]^{0.1}$$

ρ_G and ρ_L: density of gas and liquid ($kg.m^{-3}$)

μ_G and μ_L: viscosity of gas and liquid (Pa.s).

The inverse I_x of Lockart and Martinelli's coefficient X is then given by:

$$I_x = \left[\frac{x}{1-x}\right]^{0.9} / \psi$$

x: fraction of gas in the mixture, by weight

Then, take:

$$\phi^2 = 1 + 21\, I_x + I_x^2$$

And:

$$F = \left[\phi^2\right]^{0.44}$$

F is the amplifying coefficient (greater than 1) of the convection coefficient. This reflects the fact that the presence of bubbles activates heat transfer by convection.

In a tube, the convection coefficient is given by:

$$\alpha_c = 0.023 \frac{\lambda_L}{d_i} \mathrm{Re}_L^{\ 0.8} \mathrm{Pr}_L^{0.33} F = \alpha_c^{(0)} F$$

λ_L: thermal conductivity of the liquid ($W.m^{-1}.°C^{-1}$)

$$\mathrm{Re}_L = \frac{4W}{\mu_L P_T}$$

W: total liquid and vapor flow ($kg.s^{-1}$)

P_T: perimeter of tube:

$$P_T = \pi d_i$$

d_i: internal diameter of tube (m)

$$\mathrm{Pr}_L = \frac{C_L \mu_L}{\lambda_L}$$

C_L: thermal capacity of the liquid ($J.kg^{-1}.°C^{-1}$)

The heat transfer due to boiling is reduced, as, as soon as bubbles appear, they are swept away by the liquid current.

The boiling coefficient $\alpha_{eb}^{(0)}$ is calculated using Forster and Zuber's method [FOR 55] as for stagnant boiling, with the application of a limiting coefficient (of less than 1), S, to $\alpha_{eb}^{(0)}$:

$$\alpha_{eb} = \alpha_{eb}^{(0)} S$$

With:

$$S = \exp\left[\frac{-Re_L \times F^{1.25}}{10^5}\right]$$

In this expression, the value of S is over-estimated for high values of $ReF^{1.25}$, but this is not problematic, as the relative importance of boiling is low in this case.

Finally, the convective boiling coefficient is the sum of two terms:

$$\alpha_{eb\,c} = \alpha_c^{(0)} F + \alpha_{eb}^{(0)} S$$

EXAMPLE 7.4.–

$\alpha_{eb}^{(0)} = 5{,}605\ W.m^{-2}.°C^{-1}$ $\lambda_L = 0.64\ W.m^{-1}.°C^{-1}$

$C_L = 4{,}180\ J.kg^{-1}.°C^{-1}$ $d_i = 0.020\ m$

$\rho_L = 1{,}000\ kg.m^{-3}$ $\rho_G = 0.6\ kg.m^{-3}$

$\mu_L = 0.25 \times 10^{-3}\ Pa.s$ $\mu_G = 20 \times 10^{-6}\ Pa.s$

$x = 0.05$ $W = 0.3\ kg.s^{-1}$

$$\psi = \left[\frac{0.6}{1000}\right]^{0.5} \times \left[\frac{0.25}{20 \times 10^{-3}}\right]^{0.1} = 0.0315$$

$$I_x = \left[\frac{0.05}{0.95}\right]^{0.9} / 0.0315 = 2.2429$$

$$\phi^2 = 1 + 21 \times 2.2429 + 2.2429^2 = 53.13$$

$$F = 53.13^{0.44} = 5.743$$

$$Re_L = \frac{4 \times 0.3}{0.25 \times 10^{-3} \times \pi \times 0.02} = 76394$$

$$\mathrm{Pr_L} = \frac{4180 \times 0.25 \times 10^{-3}}{0.64} = 1.63$$

$$\alpha_c = 0.023 \times \frac{0.64}{0.02} \times 76394^{0.8} \times 1.63^{0.33} \times 5.743$$

$$\alpha_c = 40108 \ \mathrm{W.m^{-2}.°C^{-1}}$$

$$\mathrm{S} = \exp\left[\frac{-76394 \times 5.743^{1.25}}{10^5}\right] = 0.0011$$

$$\alpha_{ebc} = 40108 + 5605 \times 00011$$

$$\alpha_{ebc} = 40114 \ \mathrm{W.m^{-2}.°C^{-1}}$$

NOTE (Wall drying).–

When wall drying occurs, the convective boiling coefficient disappears, and is replaced by a much lower coefficient (convection + radiation), calculated as if all of the mixture were in a gaseous state.

The heat flow in the tubes varies very little in cases of drying. These cause a sudden and significant increase in the wall temperature, which can reach a level damaging to the mechanical stability of the tubes. This is known as burning, which, along with drying, constitutes a phenomenon which should be avoided.

In certain cases, cooling is carried out by direct boiling of cooling fluid. When the exchanger in question has a surface smaller than 5 m^2, total evaporation of the fluid may be required. Part of the wall therefore dries out completely, and the transfer coefficient of the device as a whole is significantly reduced. For this reason, it is better to use a larger device, evaporating only 50% of the cooling fluid. The remaining fluid is then extracted from the system and recycled.

8

Thermosiphon Reboilers

8.1. Manual method

8.1.1. *Operating principle*

Circulation in the boiler is the result of a balance between the weight of liquid held in a nearby reservoir (or held in the bottom of a distillation column) and pressure losses occurring along the trajectory of the liquid or liquid–vapor mixture (see Figure 8.1).

The liquid enters at the bottom of the boiler. It does not boil immediately, as the pressure in B is greater than the pressure in A. Between B and C, the liquid is reheated, while its pressure decreases.

The temperature at point C is equal to the boiling temperature corresponding to the pressure of the liquid, and boiling begins. From C on, a mixture of vapor and liquid is present, with the proportion of vapor increasing up to the top of the pipes.

Generally speaking, between 5 and 20% of the input liquid is evaporated; in other terms, the recirculation rate, which is the ratio between the input flow and the evaporated quantity, varies between 5 and 20.

The reservoir, or column bottom, acts as a separator between the vapor, which is evacuated, and the liquid, which is recirculated.

Between points C and D, the balance of liquid and vapor is maintained; however, the pressure is reduced. Consequently, the temperature decreases until the boiling temperature corresponding to pressure P_A, i.e. the input temperature, is reached.

In cases involving a mixture rather than a pure body, boiling reduces the proportion of light components in the liquid, and the boiler output temperature is hotter than the input temperature. Cases of this type cannot be dealt with manually, and a computer is required.

The manual method discussed below consists of taking a device and a recirculation rate, and checking that the system operates correctly.

8.1.2. *Identifying the level where boiling starts*

Let L_{PR} be the elevation of point C in relation to the lower tubular plate, and let $L_T–L_{PR}$ the vertical distance from point C to point D (point D corresponds to the upper tubular plate).

Fair provides a method for determining L_{PR}.

Consider any given point in zone BC at a temperature T and pressure P.

Fair considers that the pressure diminished in a linear manner with the elevation Z; this is true, as the pressure loss is proportional to the distance traveled. He also considers that the heating of the liquid increases in a linear manner with the elevation Z.

Using these two hypotheses, we may write:

$$\frac{T - T_B}{P_B - P} = \frac{\partial T/\partial Z}{|\partial P/\partial Z|} > 0$$

Now consider a point located between C and D. At this point, the temperature and pressure are linked by the pressure curve of saturating vapor.

$$\frac{T - T_A}{P - P_A} = \partial T/\partial \pi$$

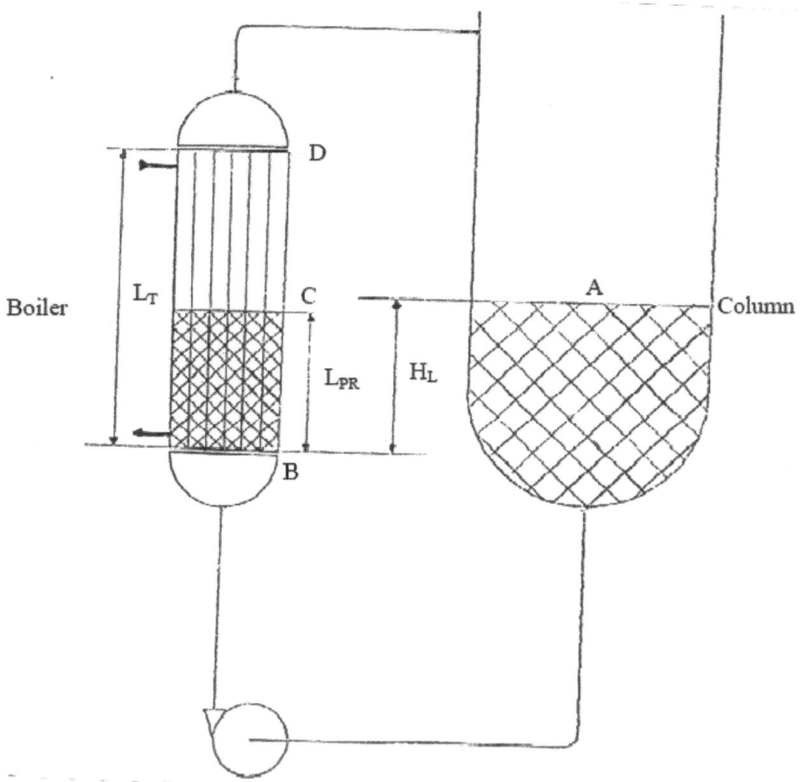

Figure 8.1. *Layout of a boiler and column assembly*

These equations are valid at point C, which is common to both zones. Furthermore, if the boiler is insulated, we may write that $T_A = T_B$.

Hence:

$$\frac{T_C - T_A}{P_B - P_C} = \frac{\partial T/\partial Z}{|\partial P/\partial Z|} \text{ and } \frac{T_C - T_A}{P_C - P_A} = \partial T/\partial \pi$$

Eliminating $T_C - T_A$:

$$(P_B - P_C)\frac{\partial T/\partial Z}{|\partial P/\partial Z|} = (P_C - P_A)\partial T/\partial \pi$$

Or:

$$\frac{P_B - P_C}{\partial T/\partial \pi} = \frac{P_C - P_A}{(\partial T/\partial Z)/|\partial P/\partial Z|} = \frac{P_B - P_A}{\dfrac{\partial T/\partial Z}{|\partial P/\partial Z|} + \dfrac{\partial T}{\partial \pi}}$$

Next, Fair [FAR 63a, FAR 63b] considers that the preheating zone is balanced by the height L_{PR} of liquid, and that the vaporization zone is balanced by the height $(L_T - L_{PR})$ of liquid in the same column bottom. This strong assumption excludes the pressure drop at the input and output points of the boiler, something which is not acceptable for a device operating in a vacuum.

Thus, according to Fair:

$$\frac{P_B - P_C}{P_B - P_A} = \frac{L_{PR}}{L_T}$$

where L_T = tube length

L_{PR} = preheating length

Finally:

$$\frac{L_{PR}}{L_T} = \frac{\partial T/\partial \pi}{\dfrac{\partial T/\partial Z}{|\partial P/\partial Z|} + \dfrac{\partial T}{\partial \pi}}$$

In this expression:

$\partial T/\partial \pi$ is obtained from tables or an Antoine curve,

$|\partial P/\partial Z|$ corresponds to a variation in hydrostatic pressure

$$|\partial P/\partial Z| = \rho_L g$$

ρ_L: density of liquid ($kg.m^{-3}$)

g: acceleration due to gravity ($9.81\ m\ s^{-2}$)

$\partial T/\partial Z$ is the result of heat transmission through the wall:

$$W_F C_{pL} \, \Delta T = U_{PR} N_T \, \pi \, D_T \Delta Z (T_V - T_A)$$

W_F: input liquid flow rate (kg s^{-1})

N_T: number of tubes

C_{pL}: specific heat of liquid (J kg^{-1}°C^{-1})

U_{PR}: overall transfer coefficient, calculated using classic methods (W m^{-2}°C^{-1})

D_T: external diameter of tubes (m)

$T_V - T_A$: temperature difference between heating steam and input flow

For a device operating at a pressure greater than or equal to atmospheric pressure, the increase in temperature between B and C is low, and an approximation consisting of assimilating the mean logarithmic temperature difference (MLTD) to $T_V - T_A$ is justified. This approximation is not appropriate for boiling in a vacuum..

Finally:

$$\frac{\partial T}{\partial Z} = \frac{U_{PR} N_T \, \pi \, D_T (T_V - T_A)}{W_F C_{pL}}$$

We shall now evaluate the different pressure drops encountered by the product.

8.1.3. *Calculation of pressure drops and verification of driving height of liquid in the reservoir or column bottom*

1) Pressure drop ΔT_1 in the reboiler inlet pipe:

This pressure drop is the result of:

Sudden contraction:

$$0.5\times\frac{\rho_L V_F^2}{2}$$

V_F: velocity of liquid in the pipe (m s^{-1})

Two right-angle bends:

$$2\times0.21\times\frac{\rho_L V_F^2}{2}$$

Friction in a pipe of length L_F:

$$f\frac{L_F}{D_F}\times\frac{\rho_L V_F^2}{2}$$

f: friction factor, obtained from curves

Sudden expansion:

$$1\times\frac{\rho_L V_F^2}{2}$$

2) Presence of a valve or circulation pump (ΔP_2):

Case of a valve:

Given C_V or a characteristic curve, we can calculate the corresponding pressure drop.

Using a centrifugal pump and given the characteristic curve, the discharge head H_p of the pump may be calculated as a function of the flow rate.

The pressure drop (negative) corresponding to the pump is:

$$\Delta P_2 = -\rho_L \; g \; H_p$$

3) Penetration of liquid into tubes (ΔP_3):

This pressure drop corresponds to sudden shrinkage. An upper limit for this loss is:

$$\Delta P_3 = 0.5 \frac{\rho_L V_L^2}{2}$$

V_L: velocity of liquid in tubes: m s^{-1}

4) Friction in tubes during pre-heating:

$$\Delta P_4 = f \frac{L_{PR}}{D_i} \frac{\rho_L V_L^2}{2}$$

with f calculated as before. d_i is the internal diameter of the tubes.

5) Static height of liquid:

$$\Delta P_5 = \rho_L \; g \; L_{PR}$$

6) Pressure drop due to acceleration:

The force corresponding to this pressure is equal to the creation of a momentum mv during a unit of time along the vaporization distance CD.

The pressure drop is:

$$\Delta P_6 = \frac{1}{A_i} \left[\text{flowrate}\,(mV) \text{ leaving at point D} - \text{flowrate}\,(mV) \text{ entering at point C} \right]$$

with A_i being the cross-section of the tubes in the boiler (m^2).

The mass entering the system per second is entirely liquid, and expressed as W_F (kg s^{-1}).

The velocity at point C is:

$$\frac{W_F}{\rho_L A_i}$$

The quantity of movement flow entering at point C is:

$$\frac{W_F^2}{\rho_L A_i}$$

The quantity of movement flow per second leaving point D corresponds to partial vaporization. With X the vapor weight fraction the mass flow rates are:

$$\text{liquid: } W_L = (1-X)W_F \quad \text{and} \quad \text{vapor: } W_V = XW_F$$

Taking R_L to be the volume fraction occupied by the liquid, then the parts of the cross-section occupied by the liquid and vapor respectively are:

$$\text{liquid: } R_L A_i \quad \text{vapor: } (1-R_L)A_i$$

The corresponding volumetric flow rates are:

$$\text{liquid: } \frac{(1-X)W_F}{\rho_L A_i} \quad \text{vapor: } \frac{XW_F}{\rho_V A_i}$$

The respective velocities are:

$$\text{liquid: } \frac{(1-X)W_F}{\rho_L R_L A_i} \quad \text{vapor: } \frac{X\,W_F}{\rho_V (1-R_L)A_i}$$

The momentum flowrate leaving point D is:

$$\frac{W_F^2}{A_i}\left[\frac{(1-X)^2}{\rho_L R_L} + \frac{X^2}{\rho_V (1-R_L)}\right]$$

Hence, finally:

$$\Delta P_6 = \frac{W_F^2}{\rho_L A_i^2}\left[\frac{(1-X)^2}{R_L} + \frac{\rho_L}{\rho_V}\frac{X^2}{(1-R_L)} - 1\right]$$

In practice, we calculate:

The physical properties coefficient

$$\psi = \left[\frac{\rho_V}{\rho_L}\right]^{0.5}\left[\frac{\mu_L}{\mu_V}\right]^{0.1}$$

The inverse I_x of the Lockart and Martinelli coefficient:

$$I_x = \left[\frac{X}{1-X}\right]^{0.9} / \psi$$

$$\text{Hence}: \ \phi^2 = 1 + 21\, I_x + I_x^2$$

$$\text{And finally}: \ R_L = \left[\frac{1}{\phi^2}\right]^{0.5}$$

These calculations are carried out using a vapor fraction X equal to that obtained at the reboiler output point: X_S.

Lockart and Martinelli's equation [LOC 49] is only valid for horizontal piping; however, using Fair's method gives correct results.

7) Static height of the liquid–vapor mixture:

R_L is recalculated using an average height $\overline{X}$ defined as:

$$\overline{X} = X_S/3$$

The average density of the mixture is thus:

$$\rho_M = \rho_L \times R_L + \rho_V (1 - R_L)$$

Giving:

$$\Delta P_7 = \rho_M g (L_T - L_{PR})$$

8) Friction between the wall and the liquid–vapor mixture:

We use:

$$X_m = \frac{2}{3} X_S$$

then calculate ϕ_M^2 as before. Coefficient ϕ_M^2 is the relationship between the pressure drop in a biphasic flow and the pressure drop in a purely liquid flow.

$$\Delta P_8 = \phi_M^2 \times \frac{L_T - L_{PR}}{L_{PR}} \times \Delta P_4$$

9) Pressure drop at bundle exit:

This pressure drop is the result of a sudden expansion:

$$\Delta P_9 = \phi^2 \times \frac{1}{L_{PR} / d_i} \Delta P_4$$

where ϕ^2 is calculated using X_S.

10) Pressure drop in output piping (ΔP_{10}):

This pressure drop is the result of:

– sudden contraction;

– a right-angle turn;

– friction in a pipe of length L_S;

– sudden expansion.

The calculation is carried out as if all of the flow were in a liquid state, using the expressions given for the input piping.

The result is then multiplied by ϕ^2, calculated using X_S.

11) Verification of the recirculation ratio:

The total pressure loss is:

$$\Delta P_T = \sum_{i=1}^{i=10} \Delta P_i$$

This pressure loss must be balanced out by the column of liquid present in the reservoir or the column bottom, and measured from the level of the lower tubular plate of the reboiler.

The driving height H_L of this column is such that:

$$\rho_L \ g \ H_L = \Delta P_T$$

8.1.4. *Verification of heat transfer intensity*

1) Liquid preheating:

For pre-heating, the heat exchange is:

$$Q_{PR} = U_{PR} N_T \pi D_T (T_V - T_A) L_{PR}$$

2) Heat exchange during boiling:

$$Q_E = U_E N_T \pi D_T (T_V - T_A)(L_T - L_{PR})$$

We need to determine the value of U_E. We shall use Chen's method, using Fair's mean values for the vapor fraction.

Coefficient F is calculated for:

$$X = 0.4 \ X_S \qquad \text{(S represents the boiler outlet)}$$

The minorating coefficient S is calculated for two values of X:

$$X = X_S$$

$$X = 0.4 \ X_S$$

The arithmetic mean of the two values of S is then selected, giving the conductance or the resistance on the boiling side.

To this resistance, we add:

– the fouling resistance (boiling side);

– the resistance of the wall;

– the resistance corresponding to the condensation of the heating steam.

NOTE.–

The expression (T_V – T_A) used for the temperature difference is optimistic. In reality, the temperature of the evaporated product passes through a maximum at point C, reducing the temperature difference.

If a more precise result is required, the following method is used:

Using a first approximation of the thermal power Q_{PR}, it is possible to calculate the heating Δt_L of the liquid, and consequently the temperature at point C:

$$Q_{PR} = W_F C_{pL} \Delta t_L$$

For each of the sections, pre-heating and boiling, the MLTD is used.

NOTE.–

Using recommended recirculation rates, the walls of the device will not dry out. This remains true if even 50% of the input is evaporates (i.e. if the recirculation ratio is reduced to 2).

Note that large exchangers *cooled by evaporation of coolant fluid operate with a recirculation rate of 2, and the walls in these systems do not dry out.*

A poorly designed or badly installed reboiler may exhibit unstable operating behavior. This problem will be addressed in section 8.3.

8.2. Computerized method

8.2.1. *Reboiler calculations through simulation*

The geometry of the system is pre-defined, specifically with regard to the level of liquid in the reservoir or column bottom.

We wish to determine the exchanged heat and the recirculation ratio, i.e. the flowrate traveling through the device.

The calculation is carried out iteratively, adjusting the flow rate from one test to the next. For the first iteration, the mass flowrate density G (and thus the flow rate) is estimated, using a velocity of 0.1 m s^{-1} in the preheating zone.

It is then possible to calculate the pressure drops.

1) Input piping:

Based on the results obtained using the manual method, the pressure drop is:

$$\Delta P_F = \frac{\rho_L V_F^2}{2}\left[0.5 + 0.42 + f\frac{L_F}{D_F} + 1\right]$$

Or:

$$\Delta P_F = \frac{\rho_L V_F^2}{2}\left[2 + f\frac{L_F}{D_F}\right]$$

f is dependent on the Reynolds number in the piping.

$$\mathrm{Re} < 2300 \qquad f = 72/\mathrm{Re}$$

$$\mathrm{Re} \geq 2300 \qquad f = 0.0175 + \frac{1.32}{\mathrm{Re}^{0.42}}$$

As the formula is homogeneous, ΔP_F is obtained in Pascals in the SI unit. This is the system used here.

Taking:

G: mass flowrate density in the tubes (kg m^{-2} s^{-1})

A_i: internal cross-section of the set of tubes (m^2)

A_F: internal cross-section of the input pipe (m^2)

We may write:

$$\Delta P_F = K_F G^2$$

With:

$$K_F = \frac{1}{2\rho_L}\left[\frac{A_i}{A_F}\right]^2\left[2+f\frac{L_F}{D_F}\right]$$

2) Presence of a pump or valve in the inlet pipe:

– for a pump:

The characteristic of the pump can be approximated using the expression:

$$H = H_p - H_1\frac{Q^2}{Q_1^2}$$

H, H_p, H_1: discharge heads (m)

Q, Q_1: flow rates (m^3 s^{-1})

The pressure loss due to the pump is negative:

$$\Delta P_p = -\rho_L\, g\, H$$

and

$$Q = GA_i/\rho_L$$

Hence:

$$\Delta P_P = K_p G^2 - \Delta P_o$$

with:

$$K_p = \frac{gH_1}{Q_1^2 \rho_L}(A_i)^2 \text{ and } \Delta P_o = \rho_L g\, H_p$$

– for a valve:

$$\Delta P_V = K_V G^2$$

with:

$$K_V = K\, A_i^2$$

(K, pressure drop coefficient of the valve).

3) Preheating zone (pressure drop expression):

This pressure drop corresponds to a sudden contraction and to friction in the tubes.

$$\Delta P_{PR} = K_{PR} G^2$$

with:

$$K_{PR} = \frac{1}{2\rho_L}\left[0.5 + f\frac{L_{PR}}{d_i}\right]$$

f: friction coefficient, calculated as before

d_i: internal diameter of tubes (m)

L_{PR}: length of pre-heating zone (m)

The length of the pre-heating zone is unknown and needs to be calculated.

We must now evaluate the pressure P_{PR} at the point where boiling begins, following the liquid from its free level in the reservoir or column bottom, where its surface is subject to the mean absolute pressure P_M.

$$P_{PR} = P_M + \rho_L g H_L - \Delta P_F - \Delta P_{PR} - \rho_L g L_{PR}$$

H_L: level of liquid in the reservoir or column, measured from the lower tubular plate (m)

i.e.:

$$P_{PR} = \left[P_M + \rho_l g H_L - \Delta P_F - \frac{0,5\ G^2}{2\rho_L} \right] - \left[\frac{fG^2}{2\rho_L d_i} + \rho_L g \right] L_{PR}$$

In other terms, P_{PR} is a decreasing linear function of L_{PR}.

i.e.:

$$P_{PR} = a - b\ L_{PR}\ (\text{pressure drop})$$

We may now express the heating of the liquid in the preheating zone:

$$W_F C_{pL} dT = K_{PRT} (T_V - T) \pi d_T N_T dL_{PR}$$

After integration, we obtain the preheating equation:

$$T_V - T_{PR} = (T_V - T_F) \exp(-X\ L_{PR})\ (\text{preheating equation})$$

In these expressions:

W_F: flow rate through the device (kg s^{-1})

U_{PRT}: overall heat transfer coefficient in the pre-heating zone

d_T: external diameter of tubes

N_T: number of tubes

T_{PR}: temperature at start of boiling (°C)

T_F: liquid input temperature (°C)

$$X = \frac{U_{PRT}\ \pi\ d_T N_T}{W_F C_{pL}}$$

The pressure P_{PR} and temperature T_{PR} are connected by an expression that accounts for the fact that the liquid (of known composition) is at boiling point.

We shall consider the case of an ideal mixture of a heavy and a light component, something which is generally true in the case of a reboiler.

Take:

x: the molar fraction of the lightweight component

β: the relative volatility of the heavy component in relation to the light component. This volatility, which is the relationship between the tension of the heavy vapor to that of the light vapor, is less than 1.

π: pressure of light vapor

Thus:

$$P_{PR} = \left[x + \beta(1-x)\right]\pi(T_{PR})$$

π will be represented using an Antoine-type formula.

Hence the following equilibrium equation:

$$P_{PR} = \left[x + \beta(1-x)\right]\exp\left(A - \frac{B}{T_{PR} + C}\right) \qquad [8.1]$$

P_{PR} is an increasing function of T_{PR}.

We thus have three equations:

– pressure drop;

– preheating;

– equilibrium.

P_{PR} and L_{PR} are eliminated from these equations in order to obtain an equation in terms of T_{PR}. In actual fact,

$$L_{PR} = (a - P_{PR})/b$$

Hence:

$$f(T_{PR}) = T_{PR} - T_V + (T_V - T_F)\exp\left[\frac{-X}{b}\left[a - \left[x + \beta(1-x)\exp\left[A - \frac{B}{T_{PR} + C}\right]\right]\right]\right] = 0$$

Function $f(T_{PR})$ is clearly an increasing function of T_{PR}. The sole root of this equation can be found using Newton's method (the tangent method).

Given T_{PR}, P_{PR} and L_{PR} may be easily deduced.

In this way, the pressure drop due to pre-heating can be calculated.

4) Boiling zone:

For calculation purposes, this zone is split into around 15 sections. From one section to the next, we need to calculate:

– the transferred heat;

– the quantity of liquid vaporized;

– the pressure variation;

– the temperature variation.

– Transferred heat:

$$\Delta Q = U_E (T_V - T)\pi d_T N_T \Delta L$$

U_E: overall transfer coefficient for boiling: $W.m^{-2}.°C^{-1}$ (calculated from a partial coefficient obtained using Chen's method): or see 7.12 and 7.21.

ΔL: length of the section (m)

$$\Delta L = \frac{L_T - L_{PR}}{N_{TR}} \qquad (N_{TR}\ \text{: number of sections})$$

– Heat balance and quantity of substance vaporized:

The mass enthalpy of the liquid/vapor mixture is:

$$I = X\left(r + C_{pV}T\right) + \left(1 - X\right)C_{pL}T$$

$$\Delta I = \frac{\Delta Q}{W_F} = \left[\frac{\partial I}{\partial X}\right]_T \Delta X + \left[\frac{\partial I}{\partial T}\right]_X \Delta T$$

$$\frac{\Delta Q}{W_F} = \left[r + \left(C_{pV} - C_{pL}\right)T\right]\Delta X + \left[C_{pL} + \left(C_{pV} - C_{pL}\right)X\right]\Delta T$$

The quantity of substance vaporized in the section is (kg s^{-1})

$$\Delta W_{VA} = W_F \Delta X = \frac{\Delta Q\left[C_{pL} + \left(C_{pV} - C_{pL}\right)X\right]\Delta T}{r + \left(C_{pV} - C_{PL}\right)T}$$

where ΔT is the temperature variation of the mixture in the section:

ΔT = output temperature-input temperature for the section.

Pressure drop:

Friction:

$$\Delta P_{FTR} = K_{FTR} \times G^2$$

with:

$$K_{FPR} = f\frac{\phi^2}{2\rho_L}\frac{\Delta L}{d_i}$$

Static head:

$$\Delta P_{STR} = \rho_M g \Delta L \qquad \text{(M for mixture)}$$

with:

$$\rho_M = \rho_L R_L + \rho_V (1 - R_L)$$

Acceleration:

$$\Delta P_{ATR} = K_{ATR} \times G^2$$

with:

$$K_{ATR} = \left[\frac{(1-X)^2}{R_L \rho_L} + \frac{X^2}{(1-R_L)\rho_V} \right]_{input}^{output}$$

Equilibrium equation and temperature variation:

Firstly, the molar fraction x of the lightweight liquid is calculated as a function of the vapor mass fraction X.

Using the definition of β:

$$\beta = \left[\frac{1-y}{1-x} \right] / \frac{y}{x} = \frac{1-y}{y} \times \frac{x}{1-x}$$

and

$$y = \frac{y_M / M_V}{y_M / M_V + (1 - y_M)/M_L} \text{ and } x = \frac{x_M / M_V}{x_M / M_V + (1 - x_M)/M_L}$$

(x_M and y_M are mass fractions).

(M_V and M_L are the molecular masses of the volatile and heavy components).

Thus:

$$\beta = \frac{1 - y_M}{y_M} \times \frac{x_M}{1 - x_M}$$

or

$$y_M = \frac{x_M}{\beta + (1-\beta)x_M}$$

Furthermore, the overall mass fraction of the lightweight volatile z_{MV} is a known quantity:

$$z_{MV} = Xy_M + (1-X)x_M = \frac{Xx_M}{\beta + (1-\beta)x_M} + (1-X)x_M$$

i.e.:

$$(1-\beta)(1-X)x_M^2 + \left[z_{MV}(1-\beta) + \beta(1-X) + X\right]x_M - \beta z_{MV} = 0$$

This is a second-degree equation in terms of x_M, and the positive root is selected. From this, we deduce:

$$x = \frac{x_M/M_V}{x_M/M_V + (1-x_M)/M_L}$$

The balance temperature corresponding to x and to pressure P is thus, following the equilibrium equation [8.1]:

$$T = \frac{B - C\left(A - \mathrm{Ln}\frac{P}{x + (1-x)\beta}\right)}{A - \mathrm{Ln}\frac{P}{x + (1-x)\beta}}$$

In this way, iterative calculations are carried out for each section. For the first iteration, the properties at the beginning of the section are used, and we calculated the vaporized quantity using $\Delta T = 0$. Averaged properties obtained using values for the beginning and end of the section are then used.

Once the whole of the vaporization zone has been treated, we calculate:

$$K_{VA} = \sum_{1}^{N_{TR}}(K_{FTR} + K_{ATR})$$

$$\Delta P_{SVA} = \sum_{1}^{N_{TR}} \Delta P_{STR}$$

5) Adiabatic expansion at the end of the bundle and in the outlet piping:

We calculate the length equivalent to this pathway:

$$L_{ES} = \frac{D_S}{f}\left[\left[\frac{A_S}{A_i}\right]^2 + 1.7 + f\frac{L_S}{D_S}\right]$$

where f is determined as if all of the liquid were liquid in the outlet piping.

The equivalent length is then divided into five sections, and calculations are carried out as for the vaporization section.

However:

– in this case, $\Delta Q = 0$

– the load loss due to the static head is null: $\Delta P_{STR} = 0$

pressure losses due to friction and acceleration are calculated using the mass flow G_S in the outlet piping. K_S, which is equivalent to K_{VA}, must be corrected by multiplying by $\left[\frac{A_i}{A_S}\right]^2$.

The recirculation flow can now be verified using the pressure loss balance:

$$P_M + \rho_L g\left(H_L + H_p\right) = \left(K_F + K_p + K_V + K_{PR} + K_{VA} + K_S\right)G^2 + \rho_L g L_{PR} + \Delta P_{SVA}$$

This expression is used to calculate the value of G for the next iteration.

The calculation process is terminated when the new value of G differs from its predecessor by less than 1%.

8.2.2. *Reboiler calculation by direct determination*

In this case, the recirculation flow rate and the vapor content of the output are predefined. The pressure on the free surface of the liquid in the reservoir is also known. It is then possible to calculate the reboiler output temperature. In this case, we need to calculate the required number of tubes N_T of a given length L_T (often 2 m).

A first estimation of the number of tubes may be obtained, taking an overall transfer coefficient of 580 W m^{-2}°C^{-1}. The diameters and length of the input piping are specified as input data.

Calculations are then carried out, starting from the outlet end. The expressions established above remain valid, but the results obtained in each section constitute the output, rather than the input, conditions for the preceding section.

The point where boiling begins is obtained automatically when X = 0.

The length of tube already covered is then added to a length corresponding to the pre-heating section, using a classic calculation and the factor.

$$\mathrm{Ln}\frac{T_V - T_F}{T_V - T_{PR}}$$

T_V: temperature of heating vapor

Let L_{TC} be the total length of tube calculated in this way. The new value for the number of tubes is deduced from this quantity:

$$N_T \times \frac{L_{TC}}{L_T}$$

The iteration process ceases when:

$$\left|1 - \frac{L_{TC}}{L_T}\right| < 1$$

The level of liquid required in the reservoir or column bottom is deduced from the pressure drop balance.

GENERAL NOTE.–

Certain products produce significant levels of fouling (tars) and leave a deposit on the wall of the tubes, reducing the cross-section available for the product; in this case, a fouling resistance must be added to the expression of transfer coefficients.

8.3. Operational stability of thermosiphon reboilers (breathing)

8.3.1. *Mechanism of the phenomenon*

Certain boilers may be seen to operate in a cyclical manner, and the pressure inside the boiler varies periodically. The boiler is said to "breathe".

We shall consider that microconvection is responsible for most of the heat exchange in this case. As the pressure in the boiler increases, the input liquid flowrate decreases, as the available pressure difference is smaller; the overall flow in the boiler also decreases. Both Chen's [CHE 66] moderating coefficient S and the transfer coefficient therefore increase.

Finally, the evaporation intensity increases, which also tends to increase the pressure, if the outlet piping is sufficiently resistant.

This cumulative phenomenon occurs very quickly, but its amplitude is limited as coefficient S cannot exceed a value of 1. The pressure therefore stabilizes at a value higher than the nominal value. During this time, the liquid, which is unable to enter the boiler due to the excess pressure in the vapor, accumulates in the column. The level of liquid rises, resulting in a higher pressure at the boiler entrance.

After around 1 min, enough liquid will have been accumulated in the column bottom, and liquid will once again enter the boiler. The evaporation intensity decreases, accelerating the liquid flow and reducing the level of liquid in the column.

Once the level in the reservoir is sufficiently low, the flow rate at the entrance to the boiler slows down, evaporation intensifies, and the cycle begins again.

8.3.2. *Practical factors affecting boiler stability*

A predominantly micro-convective regime, which is a *necessary condition for instability*, is more likely if:

1) Factor S is high, i.e. Re is low, i.e. the cross-section of the device is over-dimensioned.

2) The temperature difference between the heating fluid and the boiling fluid is high, as this results in a high difference between the temperature of the wall and the boiling fluid. The boiling coefficient in the vat becomes significant, and microconvection becomes predominant. Moreover, the temperature difference between the heating and boiling fluids features directly in the criterion. A high value of ΔT (i.e. high heat flow) can provoke instabilities, but this is not a calefaction phenomenon. The calefaction, or maximum flow may be calculated using Kutateladze's formula (see Chapter 7).

3) The flow resistance of the input fluid is low.

4) The resistance at the outlet is high. This occurs when the diameter of the outlet piping is very low.

Possible solutions include:

1) Tilting the device, bringing the vertical heights of the column inlet and outlet closer together, reducing the static head in the device.

If this modification is made without changing the level of liquid in the column relative to the boiler inlet, circulation is accelerated, promoting a macroconvective regime.

2) Reduce the difference in temperature between the heating and boiling fluids.

3) Install a valve in the boiler inlet piping.

4) Increase the diameter of the outlet piping.

NOTE.–

Certain authors have highlighted the existence of extremely rare phenomena in reboilers:

1) the geyser effect [LED 38];

2) the critical velocity effect [BER 67].

This phenomenon is not to be feared as the following velocities are generally used:

– liquid in inlet piping (1 m s^{-3});

– vapor in the device (6–10 m s^{-1});

– vapor in the output piping (10 m s^{-1}).

3) the vapor lock effect [LEE 56].

Lee's empirical correlation takes into account the physical properties and length of the tubes.

9

Concentrating Solutions by Vaporization

9.1. Introduction

9.1.1. *Terminology*

Vaporization is the operation by which a liquid is transformed into a vapor.

A liquid may be vaporized in two different ways:

1) Evaporation, where the free surface of a liquid is brought into contact with a hot, inert gas, or with overheated vapor. This operation is slow.

2) Boiling, where bubbles appear within a liquid following a reduction in pressure, or through the action of a heated surface immersed in the liquid. This method is considerably faster.

All devices using vaporization to concentrate products in the chemical and food industries operate by boiling, but the term "boiler" is used exclusively in the case of distillation techniques, and the term "vaporizer" is used in a number of different contexts. For this reason, vaporization devices are generally (and wrongly) referred to as "evaporators", and the basic vaporization operation is known as evaporation.

9.2. Boiling delay

9.2.1. *Definition*

When pure water is in a state of equilibrium with its steam, the relationship between the pressure of the vapor and the temperature of the liquid–vapor mixture is:

vapor pressure = function of the temperature

It has been noted that if the water is replaced by an aqueous solution containing a nonvolatile solute, the temperature of the solution–vapor mixture at the same pressure than before will be higher. These observations are confirmed by theoretical findings.

This higher temperature is that of the over-heated vapor. The difference between the two temperatures is known as boiling delay.

9.2.2. *Calculating boiling delay*

The boiling delay R_E is defined as:

$$R_E = T - T^*$$

T: boiling temperature of the solution (K)

T*: boiling temperature of the pure solvent at the same pressure (K)

Now, let us consider the molar enthalpy and molar entropy of the pure solvent, along with the partial molar enthalpy and partial molar entropy of the solvent in the solution. This shall be done for both the liquid and vapor phases. The pure solvent will be characterized by an asterisk.

For the pure solvent, the latent heat is:

$$\Lambda^* = h_i^{*V} - h_i^{*L}$$

The corresponding variation in entropy is:

$$s_i^{*V} - s_i^{*L} = \frac{\Lambda^*}{T^*}$$

For the solvent in the solution, the excess values will be characterized by the addition of an exponent E:

1) Enthalpies:

- vapor phase (solvent alone):

$$h_i^V = h_i^{*V} + R_E C_p^V$$

- liquid phase:

$$h_i^L = h_i^{*L} + h_i^E + R_E C_p^L$$

- effective latent heat at T:

$$\Lambda = h_i^V - h_i^L = \Lambda^* - h_i^E + R_E \left(C_p^V - C_p^L\right) \qquad [9.1]$$

2) Entropies

- vapor phase:

$$s_i^V = s_i^{*V} + \int_{T^*}^{T^*+R_E} C_p^V \frac{dT}{T}$$

- liquid phase:

$$s_i^L = s_i^{*L} + s_i^E + \int_{T^*}^{T^*+R_E} C_p^L \frac{dT}{T} - RLn\, x_i$$

- partial molar variation in entropy:

$$\frac{\Lambda}{T} = s_i^V - s_i^L = \frac{\Lambda^*}{T^*} - s_i^E + \int_{T^*}^{T^*+R_E} \left(C_p^V - C_p^L\right)\frac{dT}{T} + RLnx_i$$

Furthermore, following equation [9.1]:

$$\frac{\Lambda}{T} = \frac{\Lambda^*}{T} - \frac{h_i^E}{T} + \frac{R_E}{T}\left(C_p^V - C_p^L\right)$$

Eliminating $\frac{\Lambda}{T}$ from these final two equations, we obtain:

$$R\ln x_i + \frac{h_i^E}{T} - s_i^E = \Lambda^*\left(\frac{1}{T} - \frac{1}{T^*}\right) + \left(C_p^V - C_p^L\right)\left[(R_E/T) - \ln\left(1 + R_E/T^*\right)\right]$$

Additionally:

$$\ln\left(1 + \frac{R_E}{T^*}\right) \sim \frac{R_E}{T^*}$$

And, from chemical thermodynamics, we know that:

$$\frac{h_i^E}{T} - s_i^E = \frac{g_i^E}{T} = +RLn\gamma_i$$

Hence:

$$\frac{R_E\left(\Lambda^* + R_E\left(C_p^V - C_p^L\right)\right)}{T^*\left(T^* + R_E\right)} = RLn\gamma_i x_i$$

Take:

$$\Lambda^* + R_E\left(C_p^V - C_p^L\right) = \Lambda_T$$

Quantity Λ_T is the latent vaporization heat of the pure solvent at temperature T.

The boiling delay is therefore expressed:

$$R_E = \frac{-RT^{*2}\ln\gamma_i x_i}{\Lambda_T + RT^*\ln\gamma_i x_i}$$

Notation:

Λ_T: latent molar vaporization heat of solvent at T (J $kmol^{-1}$)

T*: boiling temperature of solvent at system pressure (K)

γ_i: given in Aseyer [ASE 98]

R: ideal gas constant (J $kmol^{-1}.K^{-1}$)

γ_i: activity coefficient of solvent in the liquid mixture

x_i: molar fraction of solvent

When calculating the fraction x_i, we consider that the ions produced by separation of the electrolyte act as separate molecules. This quantity, x_i, is very easy to calculate when the dissolved electrolyte separates entirely (as in the case of NaCl or NaOH in water). Otherwise, the electrolyte dissociation equilibrium must be taken into account.

When an electrolyte is only partly separated, the non-separated molecules are responsible for a certain level of pressure in the electrolyte vapor above the solution. In this case, evaporation – in the sense used here, where only the solvent is volatile – is not suitable as a means of increasing the level of solute in the solution, and other methods, such as distillation, should be used. In most cases, evaporation experts only need to deal with fully separated inorganic solutes. One notable exception to this rule concerns certain foodstuffs (milk, fruit juice, glucose, glucose, pulps, aqueous effluents, etc.), for which R_E is expressed using purely empirical formulas.

One example of an empirical formula is:

$$R_E = \frac{aX + b\,X^2}{1 + c\,X}$$

X: mass fraction of dry extract (total quantity of dissolved material)

Coefficients a, b and c are considered to be independent of the temperature; this is not strictly true, but the approximation errs on the side of caution as R_E is an increasing function of the temperature. Values of R_E are measured in laboratory conditions under atmospheric pressure, whereas industrial installations almost always operate in a vacuum. Coefficient c is negative, as the value of R_E increases dramatically above certain values of X.

The value of R_E is rarely in excess of 15°C for concentrates leaving a system, but there are a certain number of exceptions to this rule. For soda with a 50% fraction of NaOH, the R_E may reach around 60°C. In the case of super phosphoric acid with a 70% P_2O_5 content, the boiling delay may be of the order of 140°C.

Note that the calculation of R_E as defined above is a two-step process, as shown below.

EXAMPLE.–

Calculate the R_E of a soda containing 30% of NaOH at ambient temperature and under reduced pressure. Use a value of 0.608 for the activity coefficient of the water.

The soda is completely dissociated into two ions, hence:

$$x_i = \frac{70/18}{70/18 + 2\times 30/40} = 0.722$$

Furthermore:

$$T^* = 20°C$$

$$\Lambda^* = 43749000 \text{ J.kmol}^{-1}$$

Hence:

$$R_E^{(1)} = \frac{-8314\times 293^2 \times \text{Ln}(0.608\times 0.722)}{43749000 + 8314\times 293\times \text{Ln}(0.608\times 0.722)}$$

$$R_E^{(1)} = 14{,}1°C$$

To calculate Λ_T, we consider that:

$$T = T^* + R_E^{(1)} = 20 + 15.2 = 35.2°C$$

At 37°C:

$$\Lambda_T = 35535000 \text{ J.kmol}^{-1}$$

$$R_E = \frac{-8314 \times 293^2 \times \text{Ln}(0.608 \times 0.722)}{45535000 + 8314 \times 293 \times \text{Ln}(0.608 \times 0.722)}$$

$$R_E = 14.2°\text{C}$$

For the first stage, to calculate $R_E^{(1)}$, Λ_T has been assimilated to Λ^*.

NOTE.–

R_E may be obtained directly by measurement in a laboratory setting, and then used in manual evaporation calculations. These calculations involve the use of tables or curves featuring R_E.

The main purpose of the formulae given above is to allow calculations to be carried out using a computer.

9.3. Multiple-effect evaporation

9.3.1. *Ideal specific consumption*

First, consider the example of a single evaporator. Leaving aside the various effective heats, the flow of heating vapor condensed in the shell may be considered equal to the flow of steam produced. In other terms, the specific consumption is around 1.

(Specific consumption is equal to the consumed vapor flow consumed divided by the produced vapor flow.)

Now, imagine that the vapor produced by the first effect is used for heating purposes for a second effect.

With E_0 the live steam flow in the first effect, the flow evaporated in this effect is:

$$E_1 = E_0$$

Similarly, in the second effect:

$$E_2 = E_1$$

Giving a total evaporation:

$$E_1 + E_2 = 2E_0$$

The specific consumption is:

$$\frac{E_0}{E_1 + E_2} = \frac{1}{2}$$

In the same way, for n effects, the ideal specific consumption should be approximately $\frac{1}{n}$.

9.3.2. *Specific consumption in practice*

In a multiple-effect evaporator, variations in the effective heat of the liquor are not negligible in relation to the latent heat. In practice, specific consumption is therefore slightly higher than ideal consumption, and, for n effects:

$$\frac{E_0}{\sum_{i=1}^{n} E_i} = \frac{1.25}{n}$$

This constitutes an initial approximation, suitable for use when beginning a manual calculation of evaporation.

9.3.3. *Possible layouts*

There are two basic layouts, differing in the way the vapors and liquor circulate through the system.

Co-current circulation is the more common of the two. In this layout, the liquor is introduced into the first effect, i.e. the effect with the highest vapor pressure. Using certain types of evaporators (but not falling film evaporators), the pressure difference between the two consecutive effects is sufficient to move the liquor from one level to the next without the use of a pump. Co-current circulation is highly recommended when working with heat-sensitive products, as the liquor is in its most concentrated state when moving through the coolest part of the system (i.e. the outlet). However, this method presents certain drawbacks. Liquor tubing may become obstructed by vapor locks, as the liquid expands between one effect and the next. This must be taken into account in the choice of pipe diameters. Furthermore, the concentrated liquor is processed at a low temperature, and may therefore have high viscosity; this results in a poor heat transfer coefficient.

In a counter-current layout, the liquor is fed into the last effect at the lowest pressure, and moves toward the first effect, in the opposite direction to the vapor. In these cases, transfer pumps are required in order to increase the pressure of the liquor. Narrow diameter if tubes may be used, as expansion does not occur. Counter-current circulation is advised in cases where the liquor is highly viscous in its most concentrated state. In this case, the most viscous liquor is processed at the highest temperature, lowering the viscosity and improving heat transfer. In all cases where a high temperature is required for the concentrated liquor, the liquor must exit through the first effect.

A mixed system may also be used. Where reasons exist to use a counter-current circulation model but if the input is very hot, the substance should not be introduced via the final effect, as the expansion of the liquid would create a state of thermal unbalance in the system.

The temperatures defining the operation of an effect with index i are:

– the dew point temperature T_{Vi} of the live steam or heating vapors;

– the difference ΔT_i between T_{Vi} and the temperature T_{Li} of the liquor;

– the boiling delay R_{Ei};

– the dew point temperature T_{Bi} of the exhaust vapors.

The vapors produced in effect i are used to heat effect i + 1:

$$T_{Bi} = T_{V,\, i+1}$$

Furthermore, with experience, it is possible to estimate heat transfer coefficients U_i for each effect. The thermal power transferred in an effect i is thus:

$$Q_i = U_i \Delta T_i \; S_i$$

where S_i is the transfer surface of effect i.

We shall now consider the creation of a temperature table for an installation with n effects.

The temperature T_{Bn} is chosen based on the temperature of the water entering the condenser. Temperature T_{Vi} is that of the available live steam. Finally, the boiling delays are considered to be known.

We can therefore write:

$$\sum_{i=1}^{n} \Delta T_i = T_{Vi} - \sum_{i=1}^{n} R_{Ei} - T_{Bn}$$

An additional condition is required in order to differentiate between the values of ΔT_i contained in their sum. For example, one might:

– aim to have equal surfaces S_i for each effect in order to reduce production costs;

– aim to minimize the total surface $\sum_{i=1}^{n} S_I$, in cases where the selected material is costly.

In the first situation (equal surfaces), we may write:

$$\frac{1}{S_1} = \ldots = \frac{1}{S_i} = \ldots = \frac{1}{S_n} = \ldots = \frac{\Delta T_1}{Q_1/U_1} = \ldots = \frac{\Delta T_i}{Q_i/U_i} = \ldots = \frac{\Delta T_n}{Q_n/U_n} = \frac{\sum_{i=1}^{n} \Delta T_i}{\sum_{i=1}^{n} Q_i/U_i}$$

The error generated by taking all Q_i to be equal is often low, and finally:

$$\Delta T_i = \left[\sum_{i=1}^{n} \Delta T_i\right] \times \frac{1/U_i}{\sum_{1=1}^{n} 1/U_i}$$

In the second situation (minimum total surface), we may show that:

$$\Delta T_i = \left[\sum_{i=1}^{n} \Delta T_i\right] \times \frac{\sqrt{Q_i/U_i}}{\sum_{i=1}^{n} \sqrt{Q_i/U_i}}$$

Temperatures are then easy to determine:

$$T_{Ln} = T_{Bn} + R_{En}$$

$$T_{Vn} = T_{Ln} + \Delta T_n$$

$$T_{B,\, n-1} = T_{Vn}$$

$$T_{L,\, n-1} = T_{Bn} + R_{E,\, n-1} \text{ etc.}$$

The final element in our preheating study is to establish a temperature T_{Fi} for the liquor entering each effect i. Once this value has been obtained, a thermal balance can be established for each effect.

9.3.4. *Liquor preheating*

Different solutions are used in different industrial contexts.

1) Chemical industry:

The condensates from each effect are collected, and used either:

i) to preheat the feed in cases of co-current circulation. A separate exchanger is generally installed in order to use the condensate from each effect, and the feed input flows through these exchangers in series, from the coolest to the hottest; or

ii) to reheat intermediate liquors, in the case of counter-current circulation. Each liquor is reheated by the condensates from the next effect in the sequence.

Tubular or plate exchangers may be used.

2) Food industry:

The solutions presented above cannot be used in this case, as “dead” zones may appear in spite of regular cleaning (particularly in tubular systems). These zones provide a substrate for microbial growth. However, other solutions have been identified which make use of the effective heat of condensates. The condensate from each effect is allowed to expand through a special hole into the shell of the following effect.

Nevertheless, the use of plate exchangers, which are relatively easy to clean, allows a certain number of techniques inspired by general practice in the chemical industry to be used.

Note that the preheater method is more effective in retaining heat from condensates than the simple expansion method. Heat is transmitted to the liquor at an earlier stage in the process, and the advantages of this approach are greater the more effects there are following a given device in the system.

9.3.5. *Thermal balance of an effect*

In an effect i, exhaust gas is produced at the temperature of the liquid T_{Li}. This vapor has a dew point temperature T_{Bi}, and is thus overheated by a quantity ($T_{Li} - T_{Bi}$). The heat corresponding to the evaporation of a unit of mass of this vapor is thus:

$$\Delta H_{Bi} = L_{Bi} + C_B \left(T_{Li} - T_{Bi} \right)$$

L_{Bi}: latent heat of water vaporization (J kg^{-1})

C_B: mass thermal capacity of exhaust vapor (J kg^{-1}°C^{-1})

During the process of condensation in the following effect, the unit of mass of exhaust vapor begins by losing excess heat, and the temperature decreases from T_{Li} to T_{Bi}. It then condenses, releasing a heat L_{Bi}. Finally, the quantity of heat corresponding to both the formation and condensation of the exhaust vapor is:

$$Q_{Bi} = E_i \Delta H_{Bi}$$

E_i: flow of exhaust gases formed in effect i (kg s^{-1})

An index 0 may be assigned to the live steam in the first effect, and we may write:

$$Q_{B0} = E_0 \Delta H_{B0}$$

The liquor fed into effect i from effect j is at temperature T_{Lj} and has a flow rate F_i. This liquor will take on the temperature T_{Li} of the liquid in effect i.

The heat released by the liquor is:

$$Q_{Li} = F_i C_{Li} \left(T_{Lj} - T_{Li} \right)$$

C_{Li}: mass heat of liquid in effect i (J kg^{-1}.°C^{-1})

Quantity Q_{Li} may be positive or negative.

In cases where the condensates from effect i – 1 expand into effect i, the vapor produced is the result of the variation in sensitive heats of condensates from $T_{V,\,i-1}$ to T_{Vi} and the corresponding thermal power is:

$$Q_{Ci} = \left[\sum_{k=0}^{i-1} E_k\right] C_C \left(T_{V,\,i-1} - T_{Vi}\right)$$

C_C: mass thermal capacity of condensates (J kg^{-1}.°C^{-1})

We can now write the thermal balance of effect i:

$$Q_{B,\,i-1} + Q_{Li} + Q_{Ci} = Q_{Bi}$$

From this, we deduce the flow E_i of formed exhaust vapor:

$$E_i = \frac{Q_{B,\,i-1} + Q_{Li} + Q_{Ci}}{\Delta H_{Bi}}$$

9.3.6. *Material and thermal balances of a multiple effect system (manual method)*

The calculation involves several stages:

1) Establishing the material balance, considering that the desired evaporation is divided equally between elements.

2) Using the previous balance to deduce the composition of the liquid phases, which allows us to obtain the boiling delays.

3) Creating a temperature table.

4) Establishing the thermal balance of each effect, following the direction of the vapor, i.e. using the vapor calculated for the previous effect as heating vapor for the effect in question. For the first iteration, the live vapor flow is estimated using the following expression:

$$E_0 = \frac{1.25}{n} \sum_{i=1}^{n} E_i$$

n: effects number

E_i: flowrate of vapor produced in effect i (kg s^{-1})

5) Comparing the total vapor flow obtained with the desired flow. The live vapor flow is then corrected in proportion, and step 4 is repeated. The process is repeated until convergence is obtained (two or three calculations will be sufficient).

6) Correcting the material balance to take account of the obtained vapor flowrate, then returning to step 1.

9.3.7. *Computerized calculation of the material balance*

Multiple effect calculations are rarely carried out by computer, as arrangements vary depending on the positions of the system inlets and outlets.

In this section, we shall simply describe one possible method for determining the order of succession of effects on the liquor side.

The material balance of effect i is written:

$$F_i - E_i = F_k \qquad [9.2]$$

E_i: flow of vapors formed in effect i.

F_i and F_k are feeds into effect i and effect k, and effect k is located immediately after effect k in the direction of circulation of the liquor.

The equation shown above thus includes two indices, i and k, which are not independent. This is expressed using the coefficients $\sigma_{i,m}$, and we may write:

$$F_k = \sum_{m=1}^{n} \sigma_{i,m} F_m$$

The coefficients $\sigma_{i,m}$ are all zero, except where m = k. In this case:

$$\left[\sigma_{i,m}\right]_{m=k} = 1$$

Varying i between 1 and n, we obtain equations, featuring n unknown quantities E_i and n – 1 unknown quantities F_i. One of the feeds is known, i.e. that which supplies the system with product for processing. A further unknown quantity is the live steam flowrate E_0, giving a total of 2n unknown values. Using F_0 to denote the general power supply:

$$F_i = \alpha_i F_0 + (1-\alpha_i) F_i$$

If i is the index of the effect receiving the input for the whole system:

$$\alpha_i = 1$$

Or else:

$$\alpha_i = 0$$

By definition, the term F_k which appears in the balance equation [9.2] can never be equal to F_0. However, it may be equal to F_p, i.e. the output flow of concentrated product. This flow rate is imposed, and thus known.

F_k thus becomes:

$$F_k = \sum_{m=1}^{n} \sigma_{i,m} F_m + \beta_i F_p$$

β_i is equal to 1 if the concentrate is extracted at stage i, and zero in all other cases. Naturally, if β_i = 1, all values of $\sigma_{i,m}$ are zero for values of m varying from 1 to n and for the value of i under consideration.

Finally, the general equation of the material balance is written:

$$\alpha_i F_0 + (1-\alpha_i) F_i - E_i = \sum_{m=1}^{n} \sigma_{i,m} F_m + \beta_i F_P$$

NOTE.–

A simple relationship may be obtained by adding the balance equations element by element:

$$F_0 = F_P + \sum_{i=1}^{n} E_i$$

This relationship has not been used here.

9.3.8. *Computerized calculation of overall thermal balance*

The n thermal balance equations above may be simply adapted for computer use following the same approach used for the material balances. For given temperatures and concentrations, these n equations are linear in E_i and F_i. Added to the n material balance equations, they make up a linear system of 2n equations which can then be used to calculate the n – 1 values of F_i and the n + 1 values of E_I.

As we have seen, the boiling delay may be expressed analytically as a function of dry material content, using either a theoretical or empirical formula. Similarly, the mass thermal capacity of a solution may also be expressed as a function of this content. The linear system of 2n equations is solved several times in an iterative manner, calculating the solute content each time. For example, the mass fraction of solute in the liquor exiting an effect with index i is:

$$X_i = \frac{W_S}{F_i - E_i}$$

with W_S the flow of solute introduced into the system.

9.3.9. *Heat losses*

Whether calculations are carried out manually or by computer, inevitable heat losses must be taken into account. To do this, the consumption E_0 of live vapor is increased by 2 or 3%, depending on the scale of the installation.

9.4. Vapor recompression

9.4.1. *Thermocompression of vapor*

Thermocompressors are generally used in multiple effect-installations. Exhaust vapor is collected in one of the first (generally the first) effect and compressed, using live steam to provide a driving force. The mixture is then reinjected into the first effect. Vapor is extracted from the first effect as the mass volume of the vapor is lowest at this point, allowing savings to be made in terms of the thermocompressor. Reinjection into the first effect means that the benefits of the process are applied to all of the effects in the system.

The compression ratio of the vapors processed in this way is around 2. The quantities of live steam required for a higher rate of compression would be unacceptably high. Generally speaking, the flowrate of live steam is similar to the flowrate of extracted vapor.

In practical terms, the installation of a thermocompressor has the same effect as the installation of an additional evaporation effect.

Systems of this type are highly cost-effective as regards to investment.

9.4.2. *Mechanical vapor recompression (MVR)*

This method is generally, but not exclusively, applied to single evaporators. In the case of multiple-effect evaporators, compression takes place on the first effect, for the reasons cited in the case of thermocompressor.

The compressed vapor is overheated at the point where it leaves the machine, and additional condensates must be added for saturation purposes. Moreover, the latent heat of the vapor is lower at high pressures, meaning that, in certain cases, live steam must also be added. A detailed balance is needed in order to determine whether the added condensates alone are sufficient, or if further live steam is needed.

MVR systems require significant investment due to the cost of the compressor. Live steam consumption is almost zero, but power requirements may be in excess of one megawatt. The costs and potential benefits should be carefully weighed before a decision is made.

9.5. Description, choice and calculation of evaporators

9.5.1. *Short tube evaporators (sugar refiners)*

These evaporators are of the biphasic (liquid–vapor) thermosiphon type. The tubes are around 2 m in length, with a nominal diameter of between 40 and 80 mm. 2 m pipes are widely used for thermosiphon reboilers used in distillation columns.

The tube bundle is arranged around a central well, through which the liquid separated from the vapor returns to the base of the tubes. The diameter of the well varies between 12 and 25% of the diameter of the shell.

The liquid–vapor mixture rises in the tubes. The liquid preheating zone represents 30–40% of the height of the tubes, respecting the normal proportions for biphasic thermosiphons operating in a vacuum. The recirculation ratio, which falls between 5 and 20, is comparable to that encountered in reboilers. This ratio is between the total flowrate in a tube divided by the flowrate of vapors formed in the tube.

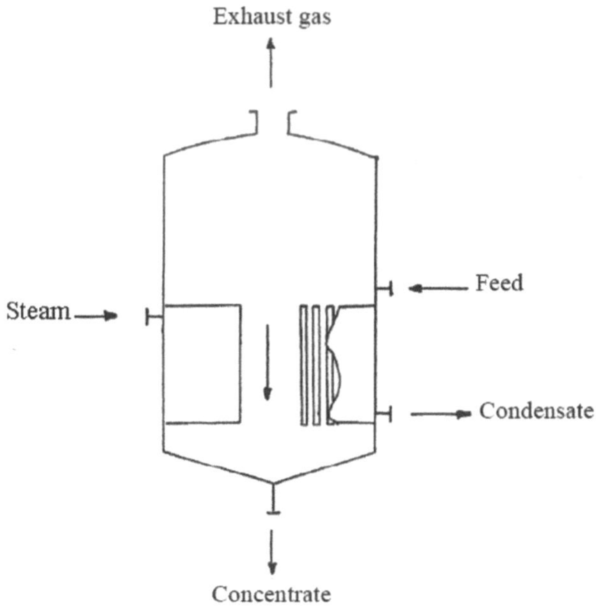

Figure 9.1. *Evaporator for sugar refining*

A manhole may be installed to enable easy cleaning of the device, making it suitable for use with lightly polluting materials. These devices are a moderate investment.

As with all devices that operate by boiling, the heat transfer coefficient is affected by the difference in temperature between the heating fluid and the processed product. The coefficient for this type of evaporator is relatively low, but still considerably greater than in boilers.

The residence time of the liquid in the device is generally a few minutes. Systems of this type should not be used for temperature-sensitive products. In terms of operational flexibility, the minimum flow rate is equal to 50% of the nominal rate.

9.5.2. *Climbing evaporators*

In this case, a solution is fed into the lower part of a bundle of vertical tubes. The liquid–vapor mixture then "climbs" up the tubes, which are around 7 m high. Vapor tends to accumulate in the axle of the tubes while the liquid remains in contact with the walls, avoiding the emergence of dry zones. The tubes have a nominal diameter of 40 mm.

The tube bundle is topped with a spherical cap deflector, which prevents vesicul liquid from escaping and renders the system suitable for processing foaming products.

The product only passes through the tubes once, and the vaporization rate is thus higher than that encountered in short-tube evaporators. Consequently, the heat transfer rate is better.

The residence time of the product is around 30 s. This type of device is not suitable for use with temperature-sensitive products. Operational flexibility is similar to that of short-tube evaporators. Finally, note that climbing evaporator systems cannot be used with products which generate fowling.

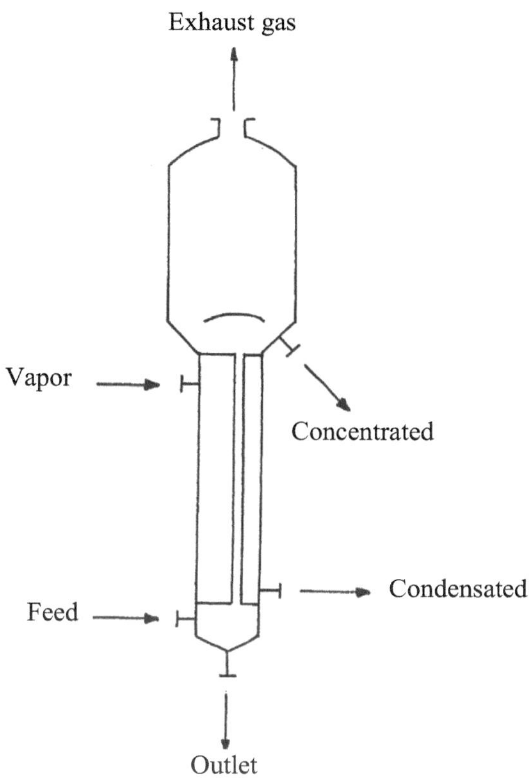

Figure 9.2. *Climbing evaporator*

9.5.3. *Forced circulation evaporators*

The heat needed for evaporation to occur is transmitted to the liquid in a separate exchanger in the form of sensible heat. The liquor then undergoes isenthalpic (free) expansion in the vapor separator. To avoid boiling in the exchanger (which would result in fouling), a head of liquid is maintained above the exchanger. The depth of this hydrostatic head is calculated from the vapor pressure of the liquor at the temperature of the tube walls with which it is in contact.

The liquor only circulates inside the tubes, and the exchanger may be installed vertically (one pass) or horizontally (two or four passes).

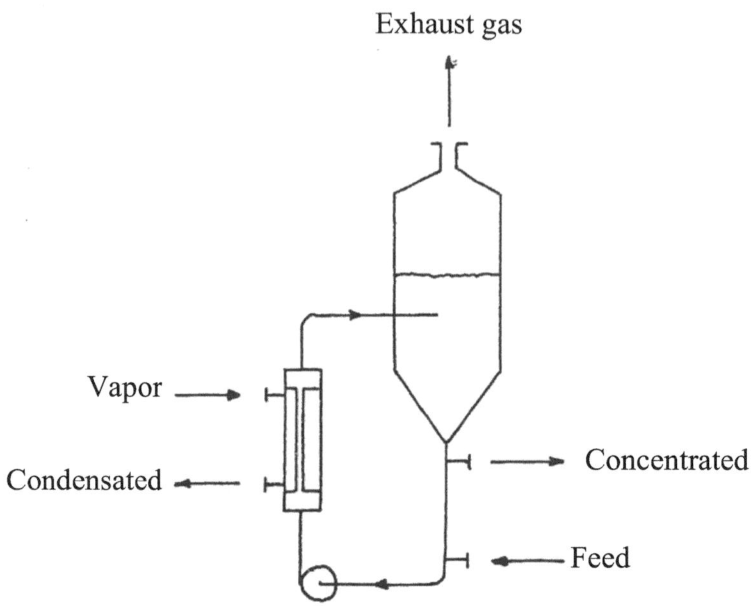

Figure 9.3. *Forced circulation evaporator*

The recirculated flow is generally calculated for a temperature increase in the exchanger of 2 or 3°C. The flow velocity through the tubes may be high (e.g. 3 $m.s^{-1}$) on the condition that the axial pump for the loop is able to compensate for the pressure loss in the circuit; in this case, precise calculations are required. The speed value given here is higher than in the case of crystallization slurry (2 $m.s^{-1}$), as there is no risk of erosion from the crystals in this case.

The use of high speeds in the tubes, without boiling, allows the treatment of products which would otherwise cause high levels of fouling. The use of a pump to maintain circulation means that highly viscous products (200 or 300 centipoises) can also be treated. Another advantage associated with high flow velocity through the tubes is that the heat transfer coefficient is considerably higher than that obtained using climbing evaporators.

Forced circulation evaporators offer a high level of flexibility, and the evaporation rate may be as low as 5% with no risk to the system; this rate may be adjusted by simply changing the flow of heating fluid to the exchanger.

The main drawback of forced circulation evaporators is their cost. A cheaper centrifugal pump may be substituted for the axial pump, although the recirculated flow will be lower. The exchange surface may also be reduced by adjusting the temperature of the heating fluid and the recirculated flowrate.

9.5.4. *Falling film evaporators*

The process liquid runs downward, by gravity, in a very thin film (a fraction of a millimeter thick) over the internal walls of the upper tubular plate of the tube bundle. The tubes are around 7 m long and the liquid is in contact with the walls for around 1 s.

Heat transfer occurs exclusively by convection, through contact between the film and the wall. Vapor bubbles do not pose a problem in this case, as they are carried off immediately by the rapid movement of the film. The transfer coefficient is therefore not dependent on the temperature difference between the wall and the liquid. This coefficient is also higher than that obtained using forced circulation systems. Falling film evaporators can be used for viscosities less than 200 centipoises.

Unlike short tube and climbing evaporators, falling film evaporators do not use a hydrostatic head to increase the boiling temperature of the liquor, and so the temperature difference between the heating fluid and process liquid is exploited to full effect.

Falling film evaporators are recommended for use with temperature-sensitive products, such as foodstuffs, and the costs involved are not excessive. Finally, products with a high potential for fouling may be processed, on the condition that the liquor is recirculated in such a way as to obtain a linear load 10 or 30 times higher than the minimum required for satisfactory damping (i.e. to avoid the emergence of dry zones).

9.5.5. *Other evaporator types*

1) Boiler evaporators:

These devices are cheap and take up relatively little space. They offer a low transfer coefficient, as the temperature difference between the heating fluid and the process liquid does not exceed 5°C, compared with 20°C for other types of evaporator. The exterior of the tubes (the surface in contact with the process liquid) is difficult to clean. As a rule, this type of evaporator is only used as a reboiler in distillation columns.

2) Scraped surface:

The product is inserted into a cylindrical shell with a heated double wall. A minimum of two scraper blades are used to ensure the renewal of the product over the wall surface the blades are helix shaped to help the product advace. This type of device is useful for highly viscous substances (up to one million centipoises), on condition that the gap between the scrapers and the wall is less than a few hundredths of a millimeter. If the gap is any bigger, the heat transfer coefficient may be reduced by a half to two-thirds.

Scraped surface evaporators are suitable for use with products which generate high levels of fouling.

The main drawback to scraped film systems is the price, which increases dramatically as the maximum gap size between the scrapers and the wall decreases.

3) Plate evaporator:

The heating surface is a set of corrugated plates separated by elastomer joints. The space between the plates varies between 0.25 and 3 mm, and the small volume available to vapors means that these devices cannot be used at absolute pressures of less than 0.4 bars. Furthermore, to protect the joints, the absolute operating pressure should not exceed 15 bar, and the temperature should not exceed 150°C.

Plate evaporators are very cost-effective, even when stainless steel, titanium or nickel-rich alloys are used. Ordinary steel should not be used in this case due to the low thickness of the plates.

9.5.6. *Evaporator calculation*

The method used for thermosiphon reboilers in distillation columns may be adapted for use with short tube and climbing evaporators.

Forced circulation exchangers are calculated using the same classic techniques as tubular exchangers.

In the case of plate evaporators, no general method exists, as the corrugated surface of the plates varies between manufacturers.

9.5.7. *Vapor separators*

All evaporators include a free space, in which droplets of liquid are separated from the vapor which carries them.

This space is cylindrical, and a free height of 2–3 m is left above the surface of the boiling liquid. The diameter of this space is calculated using the selected vapor velocity value. Reasonable results may be obtained using the following formula:

$$V_B = 0.025\sqrt{\frac{\rho_L}{\rho_V}}$$

ρ_L and ρ_V: densities of liquid and vapor (kg m^{-3})

V_B: velocity of vapor (m s^{-1})

This velocity is around one fourth of that used with knitted mesh separator pads; the diameter of these pads is therefore half that of the vapor separator.

9.5.8. *Physical properties used in the calculation of vaporization devices*

These properties are essentially dependent on solute content, particularly in the case of liquids. The use of theoretical expressions is strongly discouraged, if not impossible. Simple empirical expressions are given below:

1) Conductivity, thermal capacity, bulk density:

With ϕ the property in question, we may take:

$$\phi = \phi_{water}(1 - aX)$$

where X is the fraction by mass of solute.

2) Viscosity:

$$\mu = \frac{\mu_{water} + bX + cX^2 + dX^3}{1 + eX}$$

Coefficients b, c and d are each of the form:

$$A.\exp\left[\frac{B}{T + C}\right]$$

In other words, these coefficients, like μ_{water}, vary according to temperature in the same way as the viscosities.

Coefficient e is a linear function of the temperature.

9.5.9. *Order of magnitude of heat transfer coefficients*

For a liquid, which may be assimilated to water and in the best possible conditions (falling film), the overall coefficient may reach 2,300 W m^{-2}°C^{-1}.

However, for a liquid with a viscosity of around 300 centipoises at process temperature, the overall coefficient may be as low as 200 W m^{-2}°C^{-1}.

The partial coefficient for exhaust vapor condensation shall be considered equal to 4,000 W m^{-2}.°C^{-1}. This coefficient is not particularly high, as these vapors always contain inert elements. The partial coefficient for live steam condensation, however, may be considered equal to 8,000 W.m^{-2}.°C^{-1}.

10

Falling Film Vaporizer (Evaporator)

10.1. General points

10.1.1. *Types of falling film evaporator*

In a falling film evaporator, a thin film of liquid falls along the interior walls of vertical tubes due to the action of gravity. These tubes are heated from the outside, generally by condensing steam.

A priori, the solvent vapor formed by the system may be evacuated from the upper or lower parts of the device. Two possible configurations are shown in Figures 10.1 and 10.2.

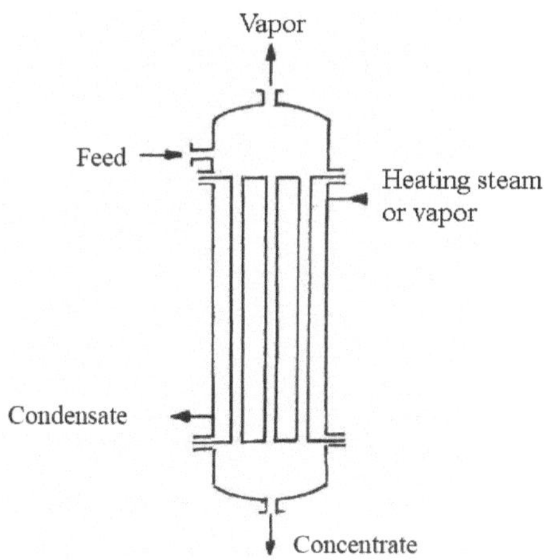

Figure 10.1. *Rising vapor*

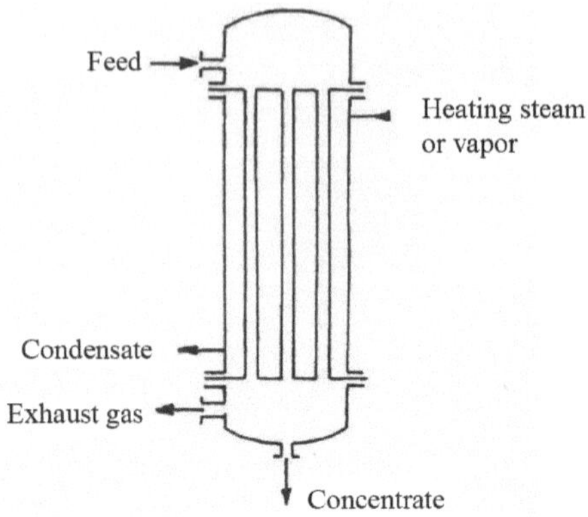

Figure 10.2. *Descending vapor*

10.1.2. *Performance*

Solutions are generally concentrated by vaporizing the solvent (generally water) with no risk to the solute, which has very low (or zero) volatility.

However, in certain cases, particularly when processing foodstuffs, the components responsible for the taste and aroma of the solution – i.e. flavors – have a non-negligible volatility. The behavior of falling film evaporators in this case will now be examined.

Devices of this type are somewhat unique in that they operate as if each elementary volume of liquid were subject to discontinuous vaporization. In classic continuous distillation, the liquid is in equilibrium with the total quantity of vapor present at each theoretical plateau. In a falling film evaporator, the liquid located in a section of thickness dZ at level Z is only in equilibrium with the mass of vapor, which it has just produced in response to heating. The average composition of the rest of the vapor in this section corresponds to the staggering of elementary values in equilibrium with sections of liquid from the end of the tube opposite the vapor outlet up to level Z.

More precisely, the evaporation of a quantity dL of water leads to:

– conversion of a quantity of solute (with index i) $y_i dL$ to the vapor phase;

– an equivalent solute reduction in the liquid phase, $d(x_i L)$.

i.e.:

$$y_i dL = x_i dL + L dx_i$$

Furthermore:

$$\frac{dL}{L} = \frac{dx_i}{y_i - x_i} = \frac{dx_i}{(E_i - 1)x_i}$$

where $E_i = y_i/x_i$ is the equilibrium coefficient of the solute.

After integration between the liquid inlet and outlet, we obtain:

$$\frac{x_{is}}{x_{ie}} = \left[\frac{L_s}{L_e}\right]^{(E_i-1)}$$

And, according to the definition of the vaporization rate v:

$$\frac{L_s}{L_e} = 1 - v$$

Hence:

$$\frac{x_{is}}{x_{ie}} = \left[\frac{1}{1-v}\right]^{(1-E_i)}$$

If the solute i is to be retained in the solution, E_i must be low; in all cases, it is less than 1, i.e. the solute must have a low volatility.

10.1.3. *Advantages of falling film evaporators*

Experience and calculations have shown that:

– the time during which the liquid is in contact with the hot tube walls is very short (around 1 s);

– heat transfer is better in film-based devices than with full tube devices;

– heat transfer operates in the same way whether boiling or simple heating is used, as vapor bubbles are immediately evacuated by the liquid film. In other terms, the void coefficient for convective boiling is zero, and only convection is involved.

Falling film evaporators are therefore suitable for use with:

– heat-sensitive products, particularly foodstuffs;

– low temperature differences between the solution and the heating fluid.

10.2. Tube wetting

10.2.1. *Minimum linear charge*

The linear charge is the part of the liquid flow in contact with the perimeter of the tubes. Ponter *et al.* [PON 67] developed an expression that gives the minimum linear charge Γ_m required for sufficient wetting of the tube surfaces, i.e. a continuous film:

$$\Gamma_m = 1.12\ \mu\left(1-\cos\theta\right)^{0.6}\left(\frac{\rho\sigma^3}{4}\right)^{0.2} \qquad [10.1]$$

Tests carried out by the authors showed that, for unpolished stainless steel (the material almost exclusively used for falling film evaporator tubes), the value of cos θ is in the region of 0.7. This value is attained after a few minutes of operation. Expression [10.1] may therefore be used with a value of:

$$\cos\theta = 0.7$$

The surface tension of pure water at 20°C is equal to 0.072 kg s^{-2}, but this value is reduced at higher temperatures and by the presence of solutes. For foodstuffs, the following value is generally used:

$$\sigma = 0.06 \text{ kg s}^{-2}$$

The viscosity (μ) is particularly important in heat transfer calculations. Expression [10.1] shows that Γ_m is proportional to μ. Falling film evaporators are only used with products with a viscosity of less than 0.3 Pa s. Above this limit, scraped surface devices should be used.

EXAMPLE 10.1.–

$\rho = 1{,}000 \text{ kg m}^{-3}$ $\mu = 0.01 \text{ Pa s}$ $g = 9.81 \text{ m s}^{-2}$

$\sigma = 0.06 \text{ kg s}^{-2}$ $\cos\theta = 0.7$

$$\Gamma_m = 1.12 \times 0.01 (1 - 0.7)^{0.6} \left[\frac{1000 \times (0.06)^3}{10^{-8} \times 9.81} \right]^{0.2}$$

$$\Gamma_m = 0.1 \text{ kg m}^{-1} \text{ s}^{-1}$$

Using a value of μ = 0.3 Pa s, without changing any of the other quantities, the linear load would be 30 times higher.

10.2.2. *Maximum linear load*

Falling film evaporators are sometimes used to concentrate products that generate high levels of fouling. The rate of fouling is reduced if the shear stress of the wall, i.e. the linear load Γ, is increased.

However, the thickness of the liquid film increases as Γ increases, and, to avoid restricting the flow of vapor in the tube, a limit of 3 kg m^{-1} s^{-1} should be respected for a viscosity of 0.10 Pa s. This corresponds to a film thickness δ of 2.5 mm. Finally, the limiting criterion for Γ is:

$$\delta = 0.0025 \text{ m}$$

As we shall see, δ increases following $\mu^{2/3}$, giving the following inequality:

$$\mu < 0.3 \text{ Pa s}$$

(presuming that the internal diameter of the tube is greater than or equal to 0.035 m).

In practice, to increase the linear load, the concentrate is partially recycled in the upper part of the tube.

10.3. Vapor velocity in the tubes

10.3.1. *Rising current for vapor*

In cases where vapor is evacuated upward, the acceptable velocity is very low in order to avoid disturbing the liquid film and blocking the device.

Diehl and Koppany developed a method for calculating the maximum velocity. This method may also be applied to vertical tube condensers used as dephlegmators, i.e. in which non-condensed elements are brought to the top, traveling in the opposite direction to the falling condensate film.

The first step in this method is to calculate the critical diameter of the tube:

$$d_c = \frac{\sigma}{3.15}$$

d_c: critical diameter (m)

σ: surface tension of liquid (N m^{-3})

Tubes used in practice generally have an internal diameter d_i greater than the critical diameter.

Next, we calculate coefficients F_1 and F_2:

If $d_i/d_c < 1$ $\quad F_1 = (d_i/d_c)^{0.4}$

If $d_i/d_c \geq 1$ $\quad F_1 = 1$

F_2 is a function of W_V/W_L at the vapor outlet point.

W_V: vapor flow (kg s^{-1})

W_L: liquid flow (kg s^{-1})

When the vapor exits and the liquid enters, $W_V/W_L = v$ and $F_2 = v^{0.25}$. Quantity v is the vaporization rate.

$$\text{First case: } F_1F_2\sqrt{\frac{\sigma}{\rho_v}} > 2.5 \qquad \left(\rho_v : \text{bulk density of the vapor (kg.m}^{-3}\text{)}\right)$$

The blockage velocity is thus:

$$V_E = 1.22 F_1 F_2 \sqrt{\frac{\sigma}{\rho_v}}$$

$$\text{Second case: } F_1 F_2 \sqrt{\frac{\sigma}{\rho_v}} < 2.5$$

$$V_E = 1.06 \left[F_1 F_2 \sqrt{\frac{\sigma}{\rho_v}} \right]^{1.15}$$

EXAMPLE 10.2.–

$\rho_V = 0.051$ kg m^{-3} (saturated water vapor at 40°C) $\quad d_i = 0.051$ m

$\sigma = 0.07$ N m^{-1} $\quad W_V/W_L = 0.2$

$$d_c = \frac{0.07}{3.15} = 0.022 \text{ m}$$

From experience, we know that *if the selected tubes have an internal diameter of at least 0.035 m, this diameter is always greater than the critical diameter*. The highest possible value of σ is 0.07, corresponding to pure water.

$$F_2 = 0.2^{0.25} = 0.67$$

$$F_1 = 1 \qquad (d_i > d_c)$$

$$F_1 F_2 \sqrt{\frac{\sigma}{\rho_v}} = 1 \times 0.67 \times \sqrt{\frac{0.07}{0.051}} = 0.78 < 2.5$$

$$V_E = 1.06 \times 0.78^{1.15} = 0.80 \text{ m.s}^{-1}$$

This velocity is considerably lower than the value of 60–100 m s^{-1} which, as we shall see, is acceptable if the vapor flows in the same direction as the falling film.

10.3.2. *Descending current for vapors*

In cases where the vapor circulates in the same direction as the liquid film, i.e. downward, a value of 20 may be used for the kinetic factor F:

$$F = V_B \sqrt{\rho_v} = 20 \text{kg}^{1/2}.\text{s}^{-1}.\text{m}^{-1/2}$$

V_B: velocity of vapor in tubes (m s^{-1})

ρ_v: vapor density (kg m^{-3})

Thus, for saturated water vapor at 40°C, i.e. at the end of a multiple effect evaporation process, we obtain:

$$V_B = \frac{20}{\sqrt{0.051}} = 90 \text{ m s}^{-1}$$

10.4. Heat transfer

10.4.1. *Partial transfer coefficient: tube side*

The method presented below was developed by Wilke [WIL 56].

The following quantities are used:

1) Reynolds number:

$$\text{Re} = \frac{4\Gamma}{\mu_L}$$

Γ: linear load of the liquid (kg m^{-1} s^{-1})

μ_L: average viscosity of the liquid (Pa s)

2) Thickness of the film:

$\text{Re} < 1600$

$$\delta = \left(\frac{3\mu_L^2 \text{Re}}{4\rho_L^2}\right)^{1/3} = 0.91\left(\frac{\mu_L^2 \text{Re}}{\rho_L^2 \text{ g}}\right)^{1/3}$$

ρ_L: bulk density of the liquid (kg m^{-3})

g: acceleration due to gravity (9.81 m s^{-2})

$$Re > 1600$$

$$\delta = \left(\frac{\mu_L^2}{\rho_L^2 \, g} \right)^{1/3} \times Re^{8/15}$$

3) Prandtl number of the liquid:

$$Pr_L = \frac{C_L \mu_L}{\lambda_L}$$

C_L: thermal capacity of the liquid (J kg^{-1}.°C^{-1})

λ_L: thermal conductivity of the liquid (W m^{-1}°C^{-1})

Depending on the value of the Reynolds number, the following relationships may be used to calculate the Nusselt number:

1) Re < 1,600 (laminar flow)

$$Nu = 0.0293 Re^{0.533} Pr^{0.344}$$

2) 1,600 < Re < 3,200

$$Nu = 0.000212 Re^{1.2} Pr^{0.344}$$

3) Re > 3,200

$$Nu = 0.00181 Re^{0.933} Pr^{0.344}$$

Any value of Nu found to be below 1.88 will be rejected and replaced by a value of 1.88.

The partial transfer coefficient α_i inside the tubes, i.e. on the film side, is written as:

$$\alpha_i = \frac{\lambda}{\delta} Nu$$

EXAMPLE 10.3.–

$\Gamma = 0.08 \text{ kg m}^{-1} \text{ s}^{-1}$ $\mu_L = 10^{-3}$ Pa s

$\rho_L = 1{,}000 \text{ kg m}^{-3}$ $C_L = 4{,}180 \text{ J kg}^{-1}\text{°C}^{-1}$

$\lambda_L = 0.46 \text{ W.m}^{-1}.\text{°C}^{-1}$

$$\text{Re} = \frac{4 \times 0.08}{10^{-3}} = 320$$

$$\delta = 0.91\left(\frac{10^{-6} \times 320}{10^{6} \times 9.81}\right)^{1/3} = 0.29 \times 10^{-3} \text{ m}$$

$$\text{Pr} = \frac{4180 \times 10^{-3}}{0.46} = 9.81$$

$$\text{Nu} = 0.0293 \times 320^{0.533} \times 9.1^{0.344}$$

$$\text{Nu} = 2.13$$

$$\alpha = \frac{0.46}{0.29 \times 10^{-3}} \times 2{,}13$$

$$\alpha = 3378 \text{W.m}^{-2}\text{°C}^{-1}$$

NOTE.–

The following table shows values of α (in $\text{W m}^{-2}\text{°C}^{-1}$) obtained by applying Wilke's method and taking extreme values of Γ and μ_L:

Γ (kg.m^{-1}.s^{-1}) / μ (Pa.s)	0.08	3
10^{-3}	3,378	7,782
0.3	432	226

For a low linear load and a relatively high viscosity, the value of coefficient α is not excessively low. This is one of the advantages of the falling film system.

10.4.2. *Overall transfer coefficient*

This coefficient is obtained by adding the series resistances:

$$\frac{1}{U}=\frac{1}{\alpha_i}+\frac{e}{\lambda}+\frac{1}{\alpha_e}$$

α_i: coefficient on film side (W m^{-2}°C^{-1})

α_e: coefficient on exterior of tubes (W m^{-2}°C^{-1})

e: thickness of tubes (m)

λ: thermal conductivity of tubes (W m^{-1}.°C^{-1})

EXAMPLE 10.4.–

$\alpha_i = 3{,}500$ W m^{-2}.°C^{-1} $\qquad$ $\alpha_e = 4{,}000$ W m^{-2}.°C^{-1}

$\lambda = 17.5$ W m^{-1}°C^{-1} $\qquad$ $e = 0.0012$ m

$$\frac{1}{U}=\frac{1}{3500}+\frac{0.0012}{17.5}+\frac{1}{4000}$$

$$=1650 \text{ W m}^{-2}.°\text{C}^{-1}$$

10.5. Distribution plate

10.5.1. *Description*

A perforated plate is placed above the uppermost tubular plate. The holes in this plate open onto the space between the tubes.

If the tubes are laid out in a triangular formation, holes will be situated on each side of the triangle and in the center of the triangle, giving 2.5 holes per half-tube, or 5 holes per tube.

The distribution plate should be covered by several centimeters of liquid in order to compensate for any horizontal tilt. A relationship exists between the flow rate and the diameter of the holes. This relationship may take one of two forms, depending on the viscosity of the liquid.

1) Viscosity of less than 50 centipoises:

The unit flow Q_u per hole is:

$$Q_u = 0.62 \frac{\pi}{4} d_o^2 \sqrt{2gh}$$

i.e.:

$$d_o = \left[\frac{4Q_u}{\pi \times 0.62 \sqrt{2gh}} \right]^{1/2}$$

EXAMPLE 10.5.–

$$Q_u = 1.32 \times 10^{-5}\ m^3.s^{-3} \qquad h = 0.04\ m$$

$$d_o = \left[\frac{4 \times 1.32 \times 10^{-5}}{\pi \times 0.62 \sqrt{2 \times 9.81 \times 0.04}} \right]^{1/2}$$

$$d_o = 0.055\ m = 0.5\ cm$$

2) Viscosity of more than 50 centipoises:

An equivalent length L_e is calculated before applying the expression of laminar flow rate in a tube.

$$L_e = d_o (0.3 + 0.04\ Re)$$

$$Re = \frac{V_o d_o \rho_L}{\mu_L}$$

$$\Delta P = \frac{128\mu_L Q_u L_e}{\pi d_o^4}$$

V_o: output velocity through the hole (m s^{-1})

ρ_L: bulk density of product (kg m^{-3})

μ_L: product viscosity (Pa s)

EXAMPLE 10.6.–

$Q_u = 10^{-5}\ m^3\ s^{-1}$ $\mu_L = 0.1$ Pa s

$d_o = 0.005$ m $\rho_L = 1{,}000$ kg m^{-3}

$$V_o = \frac{10^{-5}}{\frac{\pi}{4}\times 0.25\times 10^{-4}} = 5.12\times 10^{-1}\,m.s^{-1}$$

$$Re = \frac{0.512\times 0.005\times 1000}{0.1} = 25.6$$

$$L_e = 0.005(0.3 + 0.04\times 25.6) = 0.0067\ m$$

$$\Delta P = \frac{128\times 0.1\times 10^{-5}\times 0.0067}{\pi\times 0.0625\times 10^{-8}}$$

$$\Delta P = 436\ Pa$$

$$h = \frac{436}{1000\times 9.81} = 0.044\ m = 4.4\ cm$$

10.5.2. *Computerized calculation*

As we have seen, the overall transfer coefficient may be expressed as a function of the linear load Γ. The expression of the required tube length L_T can be shown to be a monotonically increasing function of Γ (although the coefficient α_i is not, itself, a monotonic function of the linear load, and has a minimum value). Function L_T may be inverted easily using classic numerical methods, allowing us to calculate the linear load corresponding to a given tube length L_T. In practice, a tube length of 7 m is generally used, as calculations have shown that tubes of 9 or even 12 m in length do not present a significant economic advantage.

We can then deduce the internal surface area of the evaporator:

$$S_i = \frac{L_T W_L}{\Gamma}$$

W_L: liquid load (kg s^{-1})

L_T: tube length (m)

Two equations are used to calculate the number and diameter of the tubes required.

The heat transfer surface is written as:

$$S = N_T L_T \pi d_i \qquad [10.2]$$

The cross-section through which the vapor passes is:

$$\frac{W_v}{\rho_v V_B} = N_T \frac{\pi}{4} d_i^2 \qquad [10.3]$$

Dividing [10.3] by [10.2] and rearranging the result, the internal diameter of the tubes becomes:

$$d_i = \frac{4 L_T W_V}{\rho_v V_B S}$$

And the number of tubes:

$$N_T = \frac{S}{\pi d_i L_T}$$

10.5.3. *Vapor pressure drop along the tubes*

This pressure loss is calculated using the vapor flow rate at the outlet. Considering that this rate increases from zero in a linear manner along the length of the vaporization zone, an elementary integration operation shows that a factor of 1/3 should be inserted into the pressure loss expression. Finally:

$$\Delta P = \frac{1}{3} f \frac{\rho_v V_B^2}{2} . \frac{L_v}{d_i} \qquad \text{where } f = 0.03$$

EXAMPLE 10.7.–

$$V_B \sqrt{\rho_v} = 20 \text{kg}^{1/2} \text{s}^{-1} \text{m}^{-1/2} \qquad d_i = 0.035 \text{ m} \qquad L_v = 7 \text{ m}$$

$$\Delta P = \frac{1}{3} \times 0.03 \times \frac{20^2}{2} \times \frac{7}{0.035}$$

$$\Delta P = 400 \text{ Pa} = 4 \text{ mbar}$$

NOTE.–

It is advisable to install a cyclone droplet collector to filter vapor in the lower part of the tube bundle.

10.6. Dimensioning an evaporator

10.6.1. *Tube length (preheating a cold feed)*

The corresponding tube length L_p is:

$$L_p = \frac{\Gamma C_L \delta t}{U \Delta T}$$

δt: desired preheating (°C)

ΔT: mean logarithmic temperature difference between the exterior and interior of tubes (°C)

U: overall heat transfer coefficient (W $m^{-2}.°C^{-1}$)

C_L: specific heat capacity of liquid (J $kg^{-1}.°C^{-1}$)

EXAMPLE 10.8.–

$$\Gamma = 0.21 \text{ kg s}^{-1} \text{ m}^{-1}$$

$$C_L = 2{,}508 \text{ J kg}^{-1}.°\text{C}^{-1}$$

$$\Delta t = 5°\text{C}$$

$$U = 464 \text{ W.m}^{-2}.°\text{C}^{-1}$$

$$\Delta T = 13.5°\text{C}$$

$$L_p = \frac{0.21 \times 2508 \times 5}{464 \times 13.5}$$

$$L_p = 0.42 \text{ m}$$

10.6.2. *Tube length (vaporization part)*

The total liquid load of the evaporator in the upper part of the tubes is $Q_v/\Lambda v$.

Q_v: heat transferred during vaporization (W)

Λ: latent vaporization heat (J kg^{-1})

v: vaporization ratio

The wetted perimeter is therefore:

$$\frac{Q_v}{\Lambda v} \times \frac{1}{\Gamma}$$

The heat transfer equation may be written as:

$$Q_v = \frac{Q_v}{\Lambda v} \times \frac{L_v}{\Gamma} \times \Delta T \times U$$

Hence the length of tube for this part:

$$L_v = \frac{\Gamma \Lambda v}{U \Delta T}$$

EXAMPLE 10.9.–

$\Gamma = 0.56\ \text{kg s}^{-1}\ \text{m}^{-1}$ $\qquad$ $U = 1{,}700\ \text{W m}^{-2}.°\text{C}^{-1}$

$\Lambda = 2.1 \times 10^6\ \text{J kg}^{-1}$ $\qquad$ $\Delta T = 20°\text{C}$

$v = 0.2$

$$L_v = \frac{0.56 \times 2.1 \times 10^6 \times 0.2}{1700 \times 20} = 6.9\ \text{m}$$

11

Heat Transfer in Stirred Tanks

11.1. Coil transfer

11.1.1. *General points*

When designing a reactor or crystallizer in a pilot plant, the necessary heat transfer is often carried out exclusively through the tank wall.

On an industrial scale, the transfer surface only increases as the square of the homothetic relation, while the mass of substance for processing increases as the cube of this relation. A coil must therefore be added to the tank. The heat transfer characteristics of this configuration will be studied in this chapter.

11.1.2. *Definition of the coil*

A standard coil configuration will be used:

$$\frac{D_S}{D} = 0.8 \qquad \frac{e}{d_T} = 1 \qquad \frac{H_S}{H} = 0.6$$

The diameter d_T of the coil tube represents a compromise between the utility fluid flow, resulting from the thermal balance, and the exchange surface required for heat transfer.

$$Q = \frac{\pi}{4} d_T^2 V_L \qquad S = \pi D_S \frac{H_S}{e + d_T} \pi d_T$$

If a ribbon or anchor system is used, no coil will be present, and transfer simply occurs through contact with the wall of the tank.

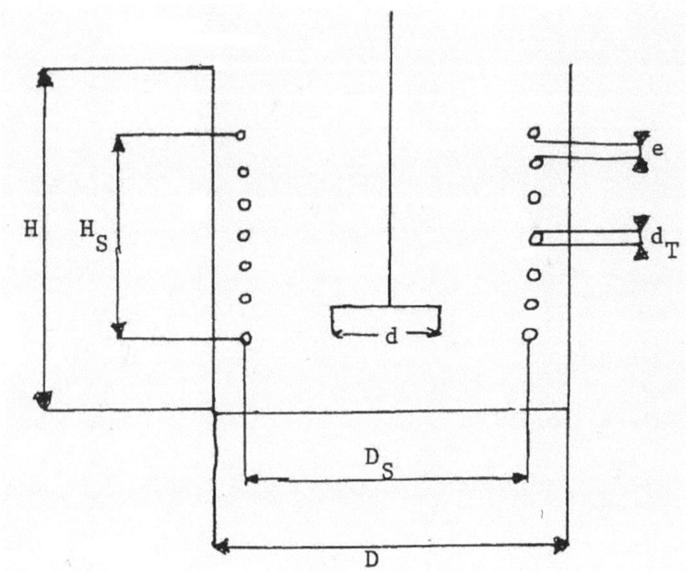

Figure 11.1. *Tank without baffles (no antivortex)*

11.2. Coefficient for the reactor wall

11.2.1. *Turbines and paddles*

The partial transfer coefficient on the wall of the tank on the stirred fluid side is given by Nagata [NAG 75]:

$$\frac{\alpha_p D}{\lambda} = 0.51\, Re^{0.66} Pr^{0.33} V_i^{0.14} \left(\frac{d}{D}\right)^{-0.25} \left(\frac{\Sigma b_i}{D}\right)^{0.15} n_p^{0.15} \left(\frac{\Sigma C_i}{H}\right)^{0.15} \sin\theta^{0.5}$$

with:

n: stirrer rotations per second

n_p: number of blades in the turbine or paddle (dimensionless)

ρ: bulk density of fluid (kg m^{-3})

μ: viscosity of liquid (Pa s)

μ_p: viscosity of liquid at temperature of transfer wall (Pa s)

V_i: viscosity coefficient:

$$V_i = \mu / \mu_p$$

Re: Reynolds number for stirring:

$$Re = \frac{d^2 n\rho}{\mu}$$

C_p: heat capacity of fluid (J kg^{-1}°C^{-1})

λ: conductivity of fluid (W m^{-1}°C^{-1})

Pr: Prandtl number:

$$Pr = \frac{C_p \mu}{\lambda}$$

i: number of impellers along the vertical shaft

b_i: vertical dimension of blades in impeller i (m)

C_i: Vertical distance of impeller i from the bottom (m)

θ: angle between blades and the horizontal plane

In order for the transfer to the tank wall to be effective, antivortex screens should not be used. This configuration will be used here.

With regard to the partial transfer coefficient on the outside of the coil, two different cases will be considered:

1) Tank without baffles with transfer across the tank wall; the impeller is positioned so that it isn't under the coil. Nagata proposes:

$$\frac{\alpha_S D}{\lambda} = 0.825\ \mathrm{Re}^{0.56}\mathrm{Pr}^{0.33}V_i^{0.14}\left(\frac{d}{D}\right)^{-0.25}\left(\frac{b}{D}\right)^{0.15} n_p^{0.15}\left(\frac{d_T}{D}\right)^{-0.3}$$

using the same notation as before.

2) Tank with baffles and without transfer via the wall, again according to Nagata:

$$\frac{\alpha_S D}{\lambda} = 2.68\ \mathrm{Re}^{0.56}\mathrm{Pr}^{0.33}V_i^{0.14}\left(\frac{d}{D}\right)^{-0.3}\left(\frac{b}{D}\right)^{0.3} n_p^{0.2}\left[\frac{C}{H}\right]^{0.75} \sin\theta^{0.5}\left(\frac{H}{D}\right)^{-0.5}$$

using the same notation as before.

11.2.2. *Marine propeller*

The tank is designed without baffles. According to Nagata, the transfer via the tank wall is obtained using:

$$\frac{\alpha_p D}{\lambda} = 0.33\ \mathrm{Re}^{0.66}\mathrm{Pr}^{0.33}V_i^{0.14}\left(\frac{d}{D}\right)^{-0.25}\left(\frac{C}{H}\right)^{0.15}$$

The transfer for the coil is obtained using:

$$\frac{\alpha_S D}{\lambda} = 1.31\ \mathrm{Re}^{0.56}\mathrm{Pr}^{0.33}V_i^{0.14}\left(\frac{d}{D}\right)^{-0.25}\left(\frac{C}{H}\right)^{0.15}$$

11.2.3. *Coil*

Taking d_i as the internal diameter of the coil tube:

$$\frac{\alpha_i d_i}{\lambda} = 0.023\left(1 + 3.5\frac{d_i}{D_S}\right)\mathrm{Re}^{0.8}\mathrm{Pr}^{1/3}$$

For very high values of D_S, the formula for rectilinear tubes may be used.

The pressure loss is given by the following expression:

$$\Delta P = 4\frac{Lf}{d_i}\frac{\rho V^2}{2} \qquad (\text{L: length of coil})$$

where f is given by White's correlation:

$$f = 0.08\ Re^{-0.25} + 0.012\left(\frac{d_i}{D_S}\right)^{0.5}$$

with:

$$Re = \frac{Vd\rho}{\mu}$$

V: velocity of the liquid (m s^{-1})

d_i: internal diameter of coil tube (m)

ρ: density of the liquid (kg m^{-3})

μ: viscosity of the liquid (Pa s)

11.2.4. *Double jacket: internal coefficient*

Two different cases exist, according to whether or not an internal spiral partition is used.

1) Double jacket with spiral liquid trajectory:

The expressions for the coil may be used in this case. The equivalent internal diameter will be considered to be equal to:

$$d_i = \frac{4A}{P}$$

where A is the cross-section through which the liquid flows and P is the full perimeter of the section used to calculate the pressure drop. In the case of heat transfer, P is simply that of the four sides of the section through which heat transfer occurs.

2) Double jacket without spiral liquid trajectory:

This arrangement is not recommended if the utility fluid is a liquid. However, if vapor is to be used to heat the double jacket, there is no advantage to be gained from using an inner spiral. In this case, a value of 6,000 $W.m^{-2}.°C^{-1}$ may be used for the partial coefficient on the vapor side.

11.3. Viscous fluids: laminar regime

11.3.1. *Ribbon*

The transfer through the wall is given by the relationship:

$$\frac{\alpha_p D}{\lambda} = 4.2\,(\mathrm{Re\,Pr})^{0.33}\,V_i^{0.2}$$

11.3.2. *Anchor*

$$\frac{\alpha_p D}{\lambda} = 1.5\,\mathrm{Re}^{0.5}\mathrm{Pr}^{0.33}V_i^{0.14}$$

11.4. Thermal conditioning of tanks and reservoirs

11.4.1. *Operating principle*

Heat is added to or removed from a container holding a given mass of liquid:

– either to compensate for undesirable heat exchange with the environment;

– or to heat or cool the mass of liquid in a given period of time.

This is carried out using a utility fluid, which may be:

– vapor, used to supply heat by condensing at a constant temperature;

– a boiling coolant liquid, which consumes heat;

– a hot or cold liquid, which creates the desired change in temperature without changing phase.

Three different configurations are generally used:

– an internal configuration with an internal exchanger (such as a coil) carrying a utility fluid;

– an external configuration with forced circulation of liquid through an external exchanger (using a pump);

– a mixed configuration, including an intermediate circulation loop, connecting an internal exchanger immersed in the liquid and an external exchanger carrying a utility fluid. This final configuration is used for emergency cooling systems in nuclear power plants.

We shall begin by describing a method for estimating the partial heat transfer coefficient for the wall surface in contact with the liquid, and then we shall give an expression of the time τ required for the temperature of the product to change from a value T_{P0} to a new value T_{P1}.

11.4.2. *Heat transfer using a coil*

Three different configurations are possible:

1) Natural convection (no stirrer):

According to Chilton *et al.* [CHI 33], the heat transfer coefficient applicable to the external surface of the coil is:

$$\alpha_e = \frac{\lambda}{d_s} 0.47 \left[\frac{d_s^3 \rho \beta g \Delta T}{\mu^2} . \frac{C_p \mu}{\lambda} \right]^{0.25}$$

d_s: diameter of the coil tube (m)

ΔT: temperature difference between the liquid and the external surface of the coil (°C)

β: volumetric expansion coefficient of the liquid($°C^{-1}$)

g: Acceleration due to gravity (m s^{-2})

C_p, μ, ρ and λ are, respectively, the thermal capacity (J kg^{-1}.°C^{-1}), the viscosity (Pa s), the bulk density (kg m^{-3}) and the thermal conductivity (W m^{-1}.°C^{-1}) of the liquid.

2) Disk turbine stirrer:

According to the formula given by Oldshue *et al.* [OLD 54]:

$$\alpha_e = \frac{\lambda}{d_s} 0.17 \left[\frac{d_I^2 N \rho}{\mu}\right]^{2/3} \left[\frac{d_I}{D}\right]^{0.1} \left[\frac{d_s}{D}\right]^{0.5}$$

d_I: turbine diameter (m)

D: tank diameter (m)

N: rotation frequency (rev s^{-1})

3) Propeller stirrer:

According to Shelland:

$$\alpha_e = \frac{\lambda}{d_s} 0.345 \left[\frac{D^2 N \rho}{\mu}\right]^{0.62} \left[\frac{D}{h}\right]^{0.27}$$

h: vertical distance between the propeller and the base of the tank (m)

d_s: diameter of coil tube (m)

D: tank diameter (m)

11.4.3. *Heat transfer via the double jacket wall*

1) No stirrer:

The wall height is almost always greater than or equal to 1 m. In this case:

$$\alpha_p = \frac{\lambda}{L_0} 0.59\ \lambda \left[\frac{L_0^3 \rho^2 g \beta \Delta T}{\mu^2} . \frac{C_p \mu}{\lambda}\right]^{0.25}$$

The homogeneity of the formula is maintained by the presence of a length L_0 equal to 1 m, whatever the height L of the wall above 1 m.

2) Disk turbine or propeller stirrer:

According to Brooks *et al.* [BRO 59] and Brown *et al.* [BRO 47]:

$$\alpha_p = \frac{\lambda}{D} 0.54 \left[\frac{D^2 N \rho}{\mu}\right]^{2/3} \left[\frac{C_p \mu}{\lambda}\right]^n \left[\frac{\mu}{\mu_p}\right]^{0.14}$$

for a disk turbine: $n = 1/3$

for a propeller: $n = 2/3$

The inclusion of four chicanes with a width of D/10 improves the heat transfer via the wall by more than 30%, on condition that the regime is turbulent, i.e.:

$$\frac{D^2 N \rho}{\mu} > 500$$

11.4.4. *Internal configuration*

For a heating process, the efficiency of the internal exchanger (double jacket or coil) relative to the utility liquid is:

$$E_{SI} = \frac{T_{S1} - T_{S2}}{T_{S1} - T_P} = 1 - e^{(-\theta_{SI})} \quad \text{with} \quad \theta_{SI} = \frac{U_I S_I}{W_S C_S}$$

T_P is the temperature of the product held in the tank.

U_I (W m^{-2} °C^{-1}) and S_I (m^2) are the overall transfer coefficient of the internal exchanger and the surface of the exchanger, respectively. θ_{SI} is the number of transfer units for the internal exchanger in relation to the service fluid. W_S and C_S are the flow rate (kg s^{-1}) and thermal capacity (J kg^{-1}°C^{-1}) of the service liquid, with input and output temperatures T_{S1} and T_{S2}.

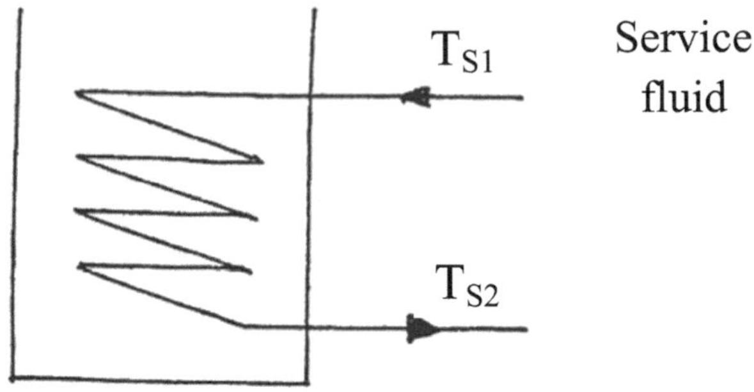

Figure 11.2. *Internal configuration*

Let Q be the instantaneous thermal power exchanged:

$$T_{S1} - T_{S2} = Q/(W_S C_S)$$

Thus:

$$M_P C_P \frac{dT_P}{d\tau} = Q = E_{SI} W_S C_S (T_{S1} - T_P)$$

M_P: mass of product contained in the tank (kg)

C_p: thermal capacity of the product (J kg^{-1}°C^{-1})

By integration, we can obtain the time τ needed to change the temperature of the product from T_{P0} to T_{P1}:

$$\tau = \frac{M_P C_P}{E_{SI} W_S C_S} \text{Ln}\left[\frac{T_{S1} - T_{P0}}{T_{S1} - T_{P1}}\right]$$

Inversely, τ can be used to deduce E_{SI}, then θ_{SI}, and, finally, the product $U_I S_I$ characterizing the exchange surface. However, for the problem to be solvable, the following conditions must be respected:

$$E_{SI} \leq 1 \quad \text{i.e. } \tau \geq \frac{M_P C_P}{W_S C_S} \text{Ln}\left[\frac{T_{S1} - T_{P0}}{T_{S1} - T_{P1}}\right]$$

In the case of cooling processes, the signs of the numerator and denominator in the logarithm are simply reversed.

11.4.5. *External configuration*

For a heating process, the efficiency of the external exchanger in relation to the product is:

$$E_{PE} = \frac{T_{PS} - T_P}{T_{S1} - T_P} = \frac{Q/(W_P C_P)}{T_{S1} - T_P} \text{ hence: } T_{S1} - T_P = \frac{Q}{E_{PE} W_P C_P}$$

W_P: product flow through the external exchanger (kg s^{-1})

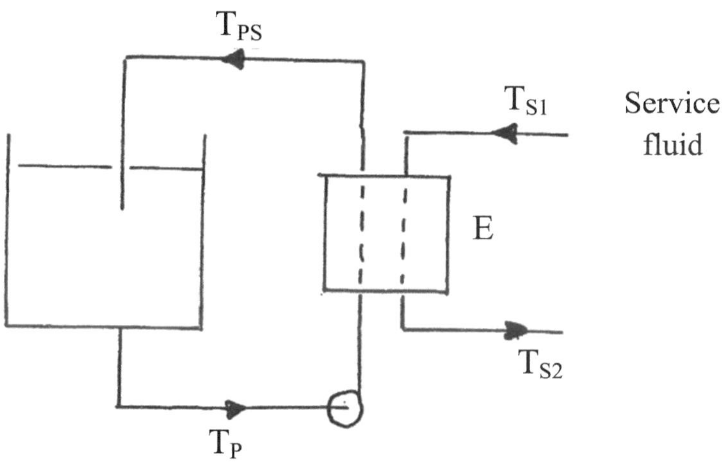

Figure 11.3. *External configuration*

Hence:

$$Q = M_P C_P \frac{dT_P}{d\tau} = E_{PE} W_P C_P (T_{S1} - T_P)$$

By integration, we can obtain the heating time τ:

$$\tau = \frac{M_P}{E_{PE} W_P} \operatorname{Ln}\left[\frac{T_{S1} - T_{P0}}{T_{S1} - T_{P1}}\right]$$

With (if the external exchanger operates using a countercurrent, as is generally the case):

$$E_{PE} = \frac{e^{\left[(1-R)\theta_{PE}\right]} - 1}{e^{(1-R)\theta_{PE}} - R}; \quad \theta_{PE} = \frac{U_E S_E}{W_P C_P}; \quad R = \frac{W_P C_P}{W_S C_S}$$

In order to calculate the product $U_E S_E$, the following conditions are necessary:

$$E_{PE} \leq 1 \text{ i.e.: } \tau \geq \frac{M_P}{W_P}\left[\frac{T_{S1} - T_{P0}}{T_{S1} - T_{P1}}\right]$$

The same expression of τ also applies to cooling of the product in the tank.

11.4.6. *Mixed configuration*

Once again, for a heating process: the efficiency of the external exchanger in relation to the liquid in the intermediate loop is:

$$E_{BE} = \frac{T_{B2} - T_{B1}}{T_{S1} - T_{B1}} = \frac{Q/(W_B C_B)}{T_{S1} - T_{B1}}$$

Thus:

$$T_{S1} - T_{B1} = \frac{Q}{E_{BE} W_B C_B} \qquad [11.1]$$

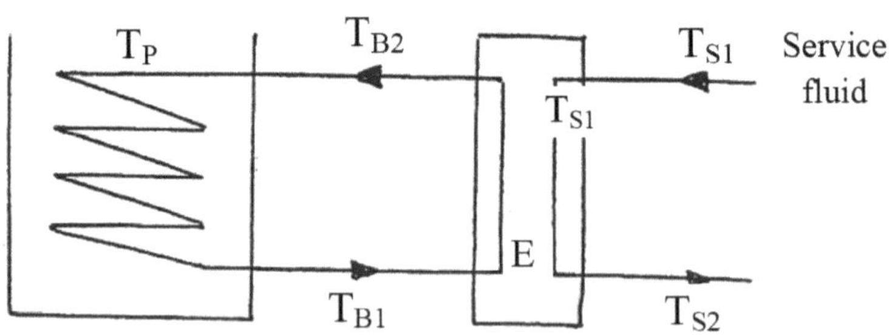

Figure 11.4. *Mixed configuration*

The efficiency of the internal exchanger, still in relation to the liquid in the loop, is:

$$E_{BI} = \frac{T_{B2} - T_{B1}}{T_{B2} - T_P} = \frac{Q/\left(W_B C_B\right)}{T_{B2} - T_P}$$

Thus:

$$T_{B2} - T_P = \frac{Q}{E_{BI} W_B C_B} \qquad [11.2]$$

Finally:

$$-\left(T_{B2} - T_{B1}\right) = \frac{-Q}{W_B C_B} \qquad [11.3]$$

Adding [11.1], [11.2] and [11.3]:

$$T_{S1} - T_P = \frac{Q}{W_B C_B}\left[\frac{1}{E_{BE}} + \frac{1}{E_{BI}} - 1\right]$$

Take:

$$\frac{1}{\xi_B} = \frac{1}{E_{BE}} + \frac{1}{E_{BI}} - 1$$

ξ_B is the overall efficiency of the system in relation to the liquid in the loop. Hence:

$$Q = M_P C_P \frac{d\tau_P}{d\tau} = \xi_B W_B C_B \left(T_{S1} - T_P\right)$$

Let us integrate:

$$\tau = \frac{M_P C_P}{\xi_B W_B C_B} \operatorname{Ln}\left[\frac{T_{S1} - T_{P0}}{T_{S1} - T_{P1}}\right]$$

with:

$$E_{BI} = 1 - e^{-\theta_{BI}} \qquad \theta_{BI} = \frac{U_I S_I}{W_B C_B}$$

$$E_{BE} = \frac{e^{[(1-R)\theta_{BE}]} - 1}{e^{[(1-R)\theta_{BE}]} - R} \qquad \theta_{BE} = \frac{U_E S_E}{W_B S_B} \qquad R = \frac{W_B C_B}{W_S C_S}$$

Indices I, E, B and S characterize the internal exchanger, the external exchanger, the loop liquid and the service fluid respectively. If the service fluid is water vapor:

$$E_{BE} = 1 - e^{-\theta_{BE}} \qquad \theta_{BE} = \frac{U_E S_E}{W_B C_B}$$

If we wish to calculate the two exchangers to obtain a given value of τ, we must begin by ensuring that:

$$\tau \geq \frac{M_P C_P}{W_B C_B} \mathrm{Ln}\left[\frac{T_{S1} - T_{P0}}{T_{S1} - T_{P1}}\right]$$

i.e.:

$$\xi_B < 1 \quad E_{BI} < 1 \quad E_{BE} < 1$$

12

Cooling or Heating of Simple-form Solids and Plant Products: Blanching

12.1. Thermal conditioning of simple-form compact solids

12.1.1. *Temperature expressions – theoretical study*

We shall consider three types of solids:

– plates, which have a small thickness in comparison to the two other dimensions;

– cylinders, which have a small diameter in comparison with their length;

– spheres.

The solid, with an initial uniform temperature t_0, is suddenly immersed in a fluid (liquid or gas) of temperature t_A, uniform and invariable over time.

Calculations are carried out using reduced temperatures.

$$\theta = \frac{t_A - t}{t_A - t_0} \quad \text{and} \quad \overline{\theta} = \frac{t_A - \overline{t}}{t_A - t_0}$$

t is the temperature at a specific point in the solid and $\overline{t}$ is the average temperature of the solid.

Furthermore, the Fourier number is involved:

$$\mathrm{Fo} = \frac{\lambda\tau}{\rho \times C_p \times R^2}$$

ρ: density of the solid (kg m^{-3})

C_p: thermal capacity of the solid (J kg^{-1}°C^{-1})

τ: time (s)

λ: thermal conductivity of the solid (W m^{-1}°C^{-1})

R: radius of the cylinder or the sphere, or half-thickness of the plate (m)

Heat transfer between the solid and its surroundings is presumed to occur by convection. The corresponding coefficient is α (W m^{-2}°C^{-1}), taken into account through the use of the Biot number:

$$\mathrm{Bi} = \frac{\alpha R}{\lambda}$$

Knowledge of the Biot number allows us to calculate the values of u_i, v_i and w_i in the following manner:

Plate:

$$\cot u_i = \frac{u_i}{\mathrm{Bi}}$$

Cylinder:

$$\frac{J_0(v_i)}{J_1(v_i)} = \frac{v_i}{\mathrm{Bi}}$$

Sphere:

$$\mathrm{tg} w_i = \frac{w_i}{\mathrm{Bi} - 1}$$

Each of these three equations has an infinite number of possible solutions, which may be numbered from 1 to infinity.

Functions J_0 (u) and J_1 (v) are zero-order Bessel functions, for which series developments may be found in a publication by Murray R. Spiegel [SPI 74].

The VDI Waermeatlas or Loncin [LON 85] may be used to obtain the first six values of u_i, v_i and w_i as a function of the Biot number.

Finally, the expressions of θ and $\overline{\theta}$ are:

Plate:

$$\theta = \sum_{i=1}^{\infty} \frac{2 \sin u_i}{u_i + \sin u_i \cos u_i} . \cos\left(u_i \frac{x}{R} \right) e^{-u_i^2 Fo}$$

$$\overline{\theta} = \sum_{i=1}^{\infty} \frac{2 \sin u_i}{u_i + \sin u_i \cos u_i} . \frac{\sin u_i}{u_i} e^{-u_i^2 Fo}$$

Cylinder:

$$\theta = \sum_{i=1}^{\infty} \frac{2 J_1(v_i)}{v_i \left[J_0(v_i)^2 + j_1(v_i)^2 \right]} . J_o\left(v_i \frac{r}{R} \right) e^{-v_i^2 Fo}$$

$$\overline{\theta} = \sum_{i=1}^{\infty} \frac{2 J_1(v_i)}{v_i \left[J_0(v_i)^2 + J_1(v_i)^2 \right]} . \frac{2 J_1(v_i)}{v_i} e^{-v_i^2 Fo}$$

Sphere:

$$\theta = \sum_{i=1}^{\infty} \frac{2(\sin w_i - w_i \cos w_i)}{w_i - \sin w_i \cos w_i} . \frac{\sin\left(w_i \frac{r}{R} \right)}{w_i \frac{r}{R}} e^{-w_i^e Fo}$$

$$\overline{\theta}=\sum_{i=1}^{\infty}\frac{2(\sin w_i - w_i\cos w_i)}{w_i - \sin w_i\cos w_i}\cdot\frac{3(\sin w_i - w_i\cos w_i)}{w_i^3}e^{-w_i^2 Fo}$$

$\overline{\theta}$ is deduced from θ using the expression:

$$\overline{\theta}=\frac{1}{V}\int_V \theta dV$$

V is the volume of the solid.

Note.–

When the solid is in the shape of a parallelepiped or a short cylinder, we may write:

Parallelepiped:

$$\theta=\theta_{width}\times\theta_{length}\times\theta_{height}$$

$$\overline{\theta}=\overline{\theta}_{width}\times\overline{\theta}_{length}\times\overline{\theta}_{height}$$

Short cylinder:

$$\theta=\theta_{cylinder}\times\theta_{length}$$

$$\overline{\theta}=\overline{\theta}_{cylinder}\times\overline{\theta}_{length}$$

EXAMPLE 12.1.–

Let us consider the *blanching* of fruit or vegetable products prior to freezing. Blanching consists of immersing a product in water at a temperature of 85–100°C or vapor at a temperature of 100°C for 1–10 min in order to destroy enzyme systems, thus reducing quality loss during subsequent storage at −18°C.

In this case, small carrots will be considered as cylinders with a diameter of 0.008 m and a length of 0.04 m. These carrots have the following properties:

$$\lambda = 0.4 \text{ W m}^{-1}.°\text{C}^{-1} \qquad C_P = 3{,}600 \text{ J kg}^{-1}°\text{C}^{-1}$$

$$\rho = 1{,}200 \text{ kg m}^{-3} \qquad t_0 = 20°\text{C}$$

The product will be immersed in water at a temperature of 90°C for a period of 100 s, and the convection coefficient is 500 W $m^{-2}°C^{-1}$.

The Fourier and Biot numbers are therefore:

Plate: $Fo = 2.31 \times 10^{-2}$ $Bi = 25$

Cylinder: $Fo = 0.5787$ $Bi = 5$

Let us calculate the first five terms of θ:

Plate:	u_i	1.46	4.40	7.40	10.30	13.30
	$\overline{\theta}_i$	0.8187	0.026	0.0078	0.0009	0
Cylinder:	v_i	1.99	4.713	7.62	10.62	13.68
	$\overline{\theta}_i$	0.089	0	0	0	0

Hence:

Plate:

$$\overline{\theta} = \sum_{i=1}^{5} \overline{\theta}_i = 0.85$$

Cylinder:

$$\overline{\theta} = \sum_{i=1}^{5} \overline{\theta}_i = 0.089$$

For the whole solid:

$$\overline{\theta} = 0.85 \times 0.089 = 0.07565 = \frac{90 - \overline{t}}{90 - 20} \text{ and so } \overline{t} = 84.7\,°C$$

The time of 100 s is almost sufficient for a temperature of 85°C to be attained, as:

$$\overline{t} = 84.7°C \# 85°C$$

12.1.2. *Auxiliary variables*

$$u = 0.860972 + 0.312133 \text{ Ln Bi} + 0.007986(\text{Ln Bi})^2 - 0.016192(\text{Ln Bi})^3 - 0.001190(\text{Ln Bi})^4 + 0.000581(\text{Ln Bi})^5$$

$$v = 1.257493 + 0.487941 \text{ Ln Bi} + 0.025322(\text{Ln Bi})^2 - 0.026568(\text{Ln Bi})^3 - 0.002888(\text{Ln Bi})^4 + 0.001078(\text{Ln Bi})^5$$

$$w = 1.573729 + 0.642906 \text{ Ln Bi} + 0.047859(\text{Ln Bi})^2 - 0.03553(\text{Ln Bi})^3 - 0.004907(\text{Ln Bi})^4 + 0.001563(\text{Ln Bi})^5$$

$$J_0(x) = 1 - \left(\frac{x}{2}\right)^2 + 0.25\left(\frac{x}{2}\right)^4 - 0.02778\left(\frac{x}{2}\right)^6 + 0.00174\left(\frac{x}{2}\right)^8$$

$$J_1(x) = \frac{x}{2} - 0.5\left(\frac{x}{2}\right)^3 + 0.8333\left(\frac{x}{2}\right)^5 - 0.00693\left(\frac{x}{2}\right)^7 + 0.00035\left(\frac{x}{2}\right)^9$$

12.2. Thermal conditioning of simple-form solids: semi-empirical study

12.2.1. *Introduction (see [CAS 84])*

In this section, we shall make use of results obtained by Castaigne, who used the fact that the relationship between the time and the local or average temperature (see below for further details) of a solid may be expressed in the following form:

$$\tau = f \,\mathrm{Ln}\left[j\left(\frac{T_1 - T_0}{T_1 - T} \right) \right]$$

Or:

$$T = T_1 + (T_0 - T_1)\, j \exp(-\tau / f)$$

T_0: initial temperature of the product (°C)

T_1: temperature of the heating fluid (°C)

T: average or local temperature of the solid (°C)

τ: contact time (s)

12.2.2. *Plates and rectangular parallelepipeds*

First, we must calculate the thermal diffusivity of the solid:

$$D = \frac{\lambda}{\rho C_p}$$

λ: thermal conductance of the solid (W m^{-1} C^{-1})

ρ: bulk density of the solid (kg m^{-3})

C: specific heat of the solid (J kg^{-1}°C^{-1})

We then calculate the Biot number:

$$Bi = \frac{\alpha L}{\lambda}$$

α: convection coefficient (W m^{-2}°C^{-1})

L: half-thickness of the plate (m)

There are three possible situations:

1) $\mathrm{Bi} \leq 0.1$

The temperature is uniform throughout the product

$$f = \frac{L^2}{\mathrm{Bi}\, D} = \frac{C_p\, L\, \rho}{\alpha} \qquad j = 1$$

2) $\mathrm{Bi} \leq 100$

$$f = 0.40528 \frac{L^2}{D}$$

The local temperature at points at a distance x from the median plane is obtained using:

$$j_x = 1.273 \cos \frac{\pi x}{2L}$$

The average temperature is obtained using:

$$j_m = 0.81104$$

3) $0.1 < \mathrm{Bi} < 100$

$$f = \frac{L^2}{D\, u^2}$$

u is given as a function of the Biot number using the formulas in section 12.1.2.

We then calculate:

$$j_c = \frac{2 \sin u}{u + \mathrm{sun}\, u \cos u}$$

And, when appropriate:

$$j_x = j_c \cos\left(\frac{u\, x}{L}\right) \qquad j_m = j_c \frac{\sin u}{u}$$

NOTE.–

In the case of a rectangular parallelepiped, f and j are given by:

$$\frac{1}{f}=\frac{1}{f_1}+\frac{1}{f_2}+\frac{1}{f_3} \qquad j=j_1 \cdot j_2 \cdot j_3$$

Indices 1, 2 and 3 characterize the three dimensions.

EXAMPLE 12.2.–

A cube of potato with sides of 2 cm and a temperature of 20°C is immersed in water at a temperature of 90°C. How long must the potato cube remain in the water to reach an average temperature of 70°C?

We use:

$$\alpha = 500 \text{ W m}^{-2}{}^{\circ}\text{C}^{-1}$$

$$\lambda = 0.5 \text{ W m}^{-1}{}^{\circ}\text{C}^{-1}$$

$$C_p = 3{,}400 \text{ J kg}^{-1}{}^{\circ}\text{C}^{-1}$$

$$\rho = 1{,}200 \text{ kg m}^{-3}$$

$$L = 0.01 \text{ m}$$

We calculate:

$$D=\frac{0.5}{3400\times1100}=1.336\times10^{-7} \text{ m}^{-2}.\text{s}^{-1}$$

$$Bi=\frac{500\times0.01}{0.5}=10$$

$$u=1.4285$$

$$f=\frac{1}{3}\times\frac{(0.01)^2}{1{,}336.10^{-7}\times(1{,}4285)^2}=122{,}3$$

$$j_c = \left[\frac{2\times 0.9899}{1.4285 + 0.9899\times 0,141817}\right]^3 = 1,2619^3 = 2,009$$

$$j_m = 2,009\times\left[\frac{0.9899}{1.4285}\right]^3 = 0.6685$$

$$\tau = 122.3\ \text{Ln}\left[0.6685\times\left(\frac{90-20}{90-70}\right)\right] = 104\ \text{s}$$

12.2.3. *Finite or infinite cylinders*

The procedure is similar to that used for the plate.

However, in this case:

$$Bi = \frac{\alpha R}{\lambda}$$

R: radius of the cylinder (m)

1) $Bi \le 0.1$

$$f = \frac{R^2}{2\ Bi\ D} = \frac{R\rho C_p}{2\alpha} \qquad j = 1$$

2) $Bi \le 100$

$$f = 0.17294\frac{R^2}{D}$$

$$j_r = 1.6015 j_o (2.4048\ r/R)$$

$$j_m = 0.6917$$

3) $0.1 < Bi < 100$

$$f = \frac{R^2}{D\ v^2}$$

$$j_c = \frac{2\,J_1(v)}{v\left[J_o^2(v) + J_1^2(v)\right]}$$

$$j_r = j_c J_o(vr/R)$$

$$j_m = \frac{2\,j_c J_1(v)}{v}$$

v is given as a function of the Biot number using the formulas in section 12.1.2.

J_0 and J_1 are zero- and first-order Bessel functions, for which series developments are provided in Appendix 1.

NOTE.–

In the case of a finite cylinder, f and j are given by:

$$\frac{1}{f} = \frac{1}{f_1} + \frac{1}{f_2} \quad \text{and} \quad j = j_1 \times j_2$$

Indices 1 and 2 refer, respectively, to an infinite cylinder of radius R and a plate with a thickness of 2 L.

12.2.4. *Sphere*

$$Bi = \frac{\alpha R}{\lambda}$$

R: radius of the sphere (m)

1) $Bi \leq 0.1$

$$f = \frac{R^2}{3\,Bi\,D} = \frac{R\rho C_p}{3\alpha} \qquad j = 1$$

2) $Bi \leq 100$

$$f = 0.10132 \frac{R^2}{D}$$

$$j_r = \frac{2 \sin(\pi r/R)}{r/R}$$

$$j_m = 0.60793$$

3) $0.1 < Bi < 100$

$$f = \frac{R^2}{d\, W^2}$$

$$j_c = \frac{2(\sin w - w \cos w)}{wr / R}$$

$$j_r = j_c \frac{\sin(wr / R)}{wr / R}$$

$$j_m = \frac{3 j_c}{w^2} \left(\frac{\sin w}{w} - \cos w \right)$$

The expression of w as a function of Bi is given in section 12.1.2.

EXAMPLE 12.3.–

Garden peas with a diameter of 8×10^{-3} m are at a temperature of 5°C. The peas are blanched by immersion into water at 90°C. How long must the peas stay in the water for the temperature in the center to reach 80°C?

We use:

$$\alpha = 500 \text{ W m}^{-2}.°\text{C}^{-1}$$

$$\lambda = 0.43 \text{ W m}^{-1}.°\text{C}^{-1}$$

$$C_p = 3{,}800 \text{ J kg}^{-1}.°\text{C}^{-1}$$

$$\rho = 1{,}100 \text{ kg m}^{-3}$$

$$R = 4 \times 10^{-3} \text{ m}$$

and calculate:

$$D = \frac{0.43}{3800 \times 1100} = 1.028 \times 10^{-7} \text{m}^2 \text{ s}^{-1}$$

$$Bi = \frac{500 \times 4.10^{-3}}{0.43} = 4.6512$$

$$w = 2{,}543$$

$$f = \frac{\left(4 \times 10^{-3}\right)^2}{1.028 \times 10^{-7} \times (2.543)^2} = 24.07$$

$$j_c = \frac{2\left[0.5634 - 2.543(-0.8262)\right]}{2.543 + 0.5634 \times 0.8262} = 1.771$$

For $R = 0$

$$j_r = j_c = 1.771$$

$$\tau = 24.07 \text{ Ln}\left[1.771 \times \left(\frac{90-5}{90-80}\right)\right] = 65 \text{ s}$$

12.3. Thermal conditioning and hydrothermal processing

12.3.1. *Roasting plant products*

Rough-cut brown tobacco (scaferlati) and coffee grains are both roasted in a system featuring a turning drum through which a hot gas, produced by burning natural gas or a sulfur-free hydrocarbon, is circulated. An internal system (using a conveyor, in the case of coffee) shakes the product whilst moving it through the system. This is known as continuous roasting.

12.3.2. *Hydrothermal processing of oilseed*

This process consists of using steam to heat a flour. The steam condenses, wetting the grains. If the humidity supplied in this way is not sufficient, water may be sprayed onto the product, using steam as a motive gas. This operation is carried out in devices known as cookers. Two types of device are available:

– Vertical cookers, which precede a batch press. This is the oldest type of system, and consists of between four and six stacked units with a double wall and double-based trays. Vapor condenses in the wall and on the base of the trays. A valve in each tray opens at regular and adjustable intervals. The grain is stirred by horizontal bars which rotate with the device shaft, and falls down to the next layer when the valve is opened. The operation takes 1–2 h. The first tank heats the grain to a temperature of 93°C, and the last to 100°C.

– Horizontal cookers, which precede a continuous press. These devices consist of fixed cylinders with a stem-heated double jacket. Inside, a shaft with oblique blades shakes and moves the product through the system. The residence time of the flour in the system is around 20 min, and the product leaves the system at a temperature of 85–95°C.

In both versions, the product is shaken for two main reasons:

– homogenization;

– to move the product through the system.

Hydrothermal processing consumes between 60 and 120 kg of steam per ton of product.

12.3.3. *Drying*

After harvesting, brown tobacco leaves are dried on trays, which are made from disjointed planks in order to ensure sufficient ventilation. Mild tobacco is dried, or cured, in drying rooms.

After hydrothermal processing, oilseed flours are dried in a device that is identical to the cooker.

Garden peas are dried on a moving bed in columns of between 4 and 8 m in height, ventilated by a hot gas.

12.3.4. *Blanching*

The blanching process consists of immersing a fruit or vegetable product in water at a temperature of 95°C for a period of between 2 and 10 min, in order to destroy enzymes capable of causing later degradation. The heat transfer involved in this process is described in section 12.1. A critical examination of various blanching procedures has been published by Philippon [PHI 84].

NOTE.–

In order to prevent further chemical changes from occurring in a plant product, it must be frozen and kept at a temperature of −18°C up until consumption. The freezing process will be discussed in Chapter 16.

13

Thermal Insulation of Piping: Tracing

13.1. Thermal insulation

13.1.1. *Insulation types*

There are four main types of insulation:

– insulation intended to protect users from contact with hot surfaces;

– insulation intended to conserve heat;

– insulation intended to keep something cold;

– cryogenic systems.

1) User protection:

The service temperature of the fluid present in a device or circulating in the pipes, along with the ambient temperature, are used to determine the thickness of insulation required in order for the external surface to remain at a temperature lower than 60°C (or 70°C in certain cases).

2) Heat conservation:

An overall heat transfer coefficient through the metallic wall and insulation material is determined as a function of the effective energy cost (or, more simply, the purchase price of crude oil). In this case, the depth of insulation required may be determined independently of temperatures.

The overall transfer coefficient generally varies between 0.5 and 0.7 W m^{-2}°C^{-1}.

When considering the thermal power of a steam boiler or a fluid-based heating system, note that the real losses in the pipes are around twice the theoretical values, due to the existence of thermal bridges created by structural support elements, valves, etc. This is also true for cooling power.

3) Cold conservation:

If the temperature of the cold fluid is higher than –20°C, the coolant system is designed in such a way that theoretical losses do not represent more than 10% of the installed cooling power. A reasonable value for the coolant heat density lost at the external surface of the coolant is 20 W m^{-2}.

If the temperature of the cold fluid is between –20°C and –40°C, theoretical losses should not represent more than 20% of the installed cooling power.

4) Cryogenic systems:

In an industrial installation systems of this type are most commonly used for distilling liquid air. This operation is carried out in two wide, short, stacked columns. The whole device is enclosed in a parallelepipedal container, known as a cold box. The space between the wall of the cold box and the equipment is filled with a granulated insulation material (such as perlite or vermiculite).

13.1.2. *Types of insulation material*

1) Cold insulation:

Cork (particularly in older installations), polyurethane or polystyrene may be used. These materials can be supplied preformed (for piping) or as sheets (for whole devices).

2) Medium-hot insulation:

Fiberglass is used at temperatures of up to 350°C, either in the form of fiber “wool”, or in felt sheets.

3) Very hot insulation:

For this procedure, fluids at temperatures of between 350°C and 550°C, mineral (or rock) wool must be used. Mineral wool is a spun silica fiber, and is used in the same way as fiber glass.

High-quality insulating materials of any form should have a thermal conductivity of less than 0.04 W/m°C and a thickness of between 40 and 130 mm is generally used.

13.1.3. *Coefficient for heat loss to the environment*

Let:

t_s: surface temperature of the device (°C)

t_a: ambient temperature (°C)

The transfer coefficient α is evaluated in W/m²°C using formulas contained in commercial documentation published by Wanner Isofi Isolation.

1) Coefficient for the inside of buildings (natural convection):

i) Flat wall:

$$\alpha_e = A(t_s - t_a)^{0.25}$$

horizontal wall (transfer in an upward direction) A = 2.15

horizontal wall (transfer in a downward direction) A = 1.13

vertical wall A = 1.59

ii) cylindrical wall:

$$\alpha_e = 1.31\left[\frac{t_s - t_a}{D}\right]^{0.25}$$

D: wall diameter (m)

2) Coefficient for the outside of buildings (forced convection):

i) flat wall:

wind speed V less than 5 m s^{-1}

$$\alpha_e = 5.22 + 3.94V$$

If V is greater than 5 m s^{-1}:

$$\alpha_e = 7.1V^{0.78}$$

ii) Cylindrical wall:

$$\alpha_e = 4.15\frac{V^{0.8}}{D^{0.2}}$$

D: wall diameter (m)

13.1.4. *Heat flow density*

Flat wall:

$$\varphi = \frac{t_i - t_a}{\frac{1}{\alpha_i} + \Sigma\frac{e_j}{\lambda_j} + \frac{1}{\alpha_e}} \qquad \left(\text{Watt m}^{-2}\right)$$

e_j and λ_j are the thickness and conductivity of a layer noted j.

α_i is the transfer coefficient between the internal liquid and the wall.

Cylindrical wall:

$$\varphi = \frac{\pi\left(t_i - t_a\right)}{\frac{1}{\alpha_i d_i} + \Sigma\frac{1}{2\lambda_j}\text{Ln}\frac{d_{j+1}}{d_j} + \frac{1}{\alpha_e d_e}} \qquad \left(\text{W par meter of length}\right)$$

A material noted k is present between two cylindrical surfaces, of diameters d_j and d_{j+1}.

13.1.5. *Temperature of the wall in contact with the atmosphere*

The heat flow density for the external surface is:

$$\varphi = U(t_i - t_a) = \alpha_e(t_p - t_a)$$

This expression may be used to calculate t_a, which is required in order to obtain a coefficient α_e for devices located inside buildings. An iterative calculation process is required, as the overall coefficient U is, itself, dependent on α_e:

For a flat wall:

$$\frac{1}{U} = \frac{1}{\alpha_i} + \sum_j \frac{e_j}{\lambda_j} + \frac{1}{\alpha_e}$$

For a cylindrical wall:

$$\frac{1}{U} = \frac{d_e}{\alpha_i d_i} + d_e \sum_j \frac{1}{2\lambda_j} \mathrm{Ln}\frac{d_{j+1}}{d_j} + \frac{1}{\alpha_e}$$

In these expressions, the overall coefficient U refers to the external surface of the insulation.

13.1.6. *Calculating the thickness of insulation*

The thermal power (in Watt) lost per m^2 of external surface of the insulation is:

$$\varphi = U(t_p - t_a) \qquad (W.m^{-2})$$

t_p and t_a are the temperature of the process fluid and the temperature of the surrounding environment, respectively.

If the purchase price of a barrel of crude oil is \$100, then:

$$U = 0.6\ W.m^{-2}.°C^{-1}$$

This value is replaced by 0.7 if the price is reduced to \$75 (in 2013).

EXAMPLE 13.1.–

Calculate the thickness of insulation required for a DN100 pipe and a 2 mm thickness of mild steel. The energy price is average:

Wind speed: $V = 5\ m.s^{-1}$

Coefficient on the process fluid side: $c_i = 2{,}000\ W\ m^{-2}.°C^{-1}$

Thermal conductivity of the insulation: $\lambda_c = 0.04\ W\ m^{-1}°C^{-1}$

Thermal conductivity of the steel: $\lambda_a = 50\ W\ m^{-1}.°C^{-1}$

Thickness of the steel: $e_a = 0.002\ m$

Let us consider a desired thickness of 0.05 m.

$$d_e = 0.10 + 0.10 = 0.2\ m$$

$$\alpha_e = 4.15 \times \frac{5^{0.8}}{0.2^{0.2}} = 20.6\ W\ m^{-2}\ °C^{-1}$$

$$\frac{1}{U} = \frac{0.20}{2000 \times 0.1} + \frac{0.002}{50} + \frac{0.20}{2 \times 0.04} \ln\left[\frac{0.20}{0.10}\right] + \frac{1}{20.6}$$

$U = 0.56\ W\ m^{-2}\ °C^{-1}$, so our choice was correct.

We see that:

– a thickness of 0.03 m corresponds to $U = 1\ W\ m^{-2}°C^{-1}$

– a thickness of 0.07 m corresponds to $U = 0.4\ W\ m^{-2}°C^{-1}$

13.2. Pipe tracing

13.2.1. *Principle*

A low-diameter heating pipe is installed along the length of a process pipe in order to prevent the process fluid from cooling down.

13.2.2. *Heat transfer coefficient*

Process pipes are usually placed in a horizontal, rather than a vertical, position. The expression of the natural convection coefficient around a horizontal pipe can therefore be used. This expression is:

$$\alpha_H = \frac{0.59\lambda}{D}\left[\frac{D^3\rho^2 g\beta\Delta T}{\mu^2}\right]^{0.25}\left[\frac{C_p\mu}{\lambda}\right]^{0.25} \qquad [13.1]$$

$C_p\ \mu/\lambda$ is the Prandtl number of air, and has a value of 0.7.

Expression [13.1] was given by Loncin [LON 69], and constitutes equation [1.135] in the cited work. The result was established by Nusselt. A number of other expressions have since been proposed.

Here, the physical properties of air at 40°C and 1 atm will be used, i.e.:

$\rho = 1.125\ \text{kg.m}^{-3}$ $\qquad \mu = 18\times 10^{-6}\ \text{Pa s}$

$\lambda = 0.031\ \text{W.m}^{-1}.°\text{C}^{-1}$ $\qquad \beta = 1/273$

Hence:

$$\alpha_H = 1.82\left(\frac{\Delta T}{D}\right)^{0.25}$$

Given that the free convection movement of the air is limited by the presence of insulation, the value used for α in this case is only 70% of the theoretical value established above, i.e.:

$$\alpha_H = 1.3\left(\frac{\Delta T}{D}\right)^{0.25} \qquad [13.2]$$

D: external diameter of tracing tube (m)

α: heat transfer coefficient (W m^{-2}°C^{-1})

ΔT is the difference in temperature between the outside wall of the tracer pipe and the air trapped between the process tube and the insulation. In a well-designed tracing system, the process fluid will exchange no heat at all with this air, which will therefore be at the same temperature as the process fluid. The heat resistance of the metallic wall of the tracer pipe will be considered negligible, and the skin temperature of the pipe will thus be considered identical to that of the tracing fluid. Finally:

$$\Delta T = (\text{tracer fluid temperature}) - (\text{process fluid temperature})$$

However, the relationship between the theoretical transfer coefficients around a vertical pipe of height H and around a horizontal pipe of diameter D is:

$$\frac{\alpha_V}{\alpha_H} = \left(\frac{D}{H}\right)^{0.25}$$

H is the height of the vertical pipe, and, if H = 3 m and D = 0.015 m, then:

$$\frac{\alpha_V}{\alpha_H} = 0.3$$

Considering the horizontal lengths to be 10 times higher than the vertical lengths, we obtain the following average coefficient:

$$1.3\left(\frac{10+1\times 0.3}{11}\right)\left(\frac{\Delta T}{D}\right)^{0.25}$$

$$\bar{\alpha} = 1.2(\Delta T/D)^{0.25}$$

EXAMPLE 13.2.–

Trace a DN100 pipe under the following conditions:

$t_p = 15°C \qquad U = 0.6\ W\ m^{-2}°C^{-1}$

$t_a = -20°C$ thickness of insulation: 0.05 m

The heat flow density lost is:

$$\varphi = 0.6(15+20) = 21\ W\ m^{-2}$$

And, per meter of tubing, taking account of thermal bridges:

$$q = 21 \times \pi \times 0.2 = 13.2\ W\ m^{-1}$$

The external diameter of the tracing pipes is 15 mm, and they transport water at an *average* temperature of 70°C.

$$\overline{\alpha} = 1.2\left[\frac{70-15}{0.015}\right]^{0.25} = 9.3\ W.m^{-2}.°C^{-1}$$

A coefficient of 1.2 has been used rather than the value of 1.3 given in equation [13.2] to take account of the fact that convection is hindered.

The power supplied per meter of tracer will be:

$$q' = 9.3 \times \pi \times 0.015 \times (70-15)$$

$$q' = 7.6\ W.m^{-2}$$

Four tracer pipes will therefore be needed, giving:

$$7.6 \times 4 = 30.4 > 26.4$$

If water flows through the tracer, with an internal diameter of 0.01 m, at a rate of 0.1 m s^{-1}, then the flow rate W will be such that:

$$1000 \times 0.1 \times (\pi / 4) \times (0.01)^2 = 0.00785\ kg.s^{-1}$$

Cooling this flow from 80°C to 60°C requires the following power:

$$0.00785 \times 4180 \times (80 - 60) = 656 \text{ W}$$

The traceable length of piping will be:

$$656/7.6 = 86 \text{ m}$$

The corresponding pressure loss is:

$$\Delta P = 0.04 \times \frac{1000 \times 0.1^2}{2} \times \frac{86}{0.015} = 1146 \text{ Pa}$$

The pressure loss in the tracer is low.

13.2.3. *Practical data*

The tracing fluid may be low-pressure vapor or water. In the latter case, the water used is demineralized, but not deoxygenated, and has the same acidity level as water in a boiler. The temperature of the tracing water rarely exceeds 80°C, considerably lower than the temperature in a boiler.

Tracer pipes have a nominal diameter of between 10 and 20 mm. As these pipes are used for heating purposes, they are located below the traced pipe, and are separated from the heat-transporting fluid by a V-shaped metal sheet. The interstitial air, heated by the tracer, rises due to natural convection, entering into contact with the wall of the process pipe.

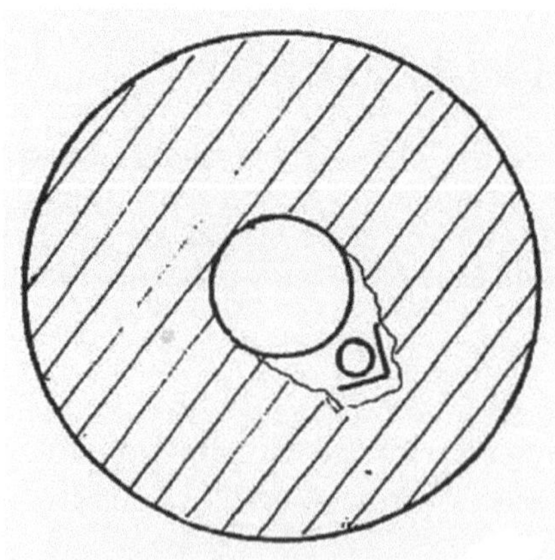

Figure 13.1. *Insulated and traced pipe*

A water tracing circuit includes the following elements (starting from the pump station):

– a water-heating exchanger;

– a venting point at a high point (low pressure) and in areas where the temperature is high. This pipe may be connected to the expansion tank;

– the tracer input collector;

– a connection to the expansion tank, which is a vertical dead-end,

– the return connection for the normal-backup pairing for the circulation pumps.

The thermal power per meter of tracer must increase if the ambient temperature decreases. This leads to an increase in the water flow rate and to a loss of pressure. The pumping power therefore increases as the external temperature decreases.

Expansion tanks are generally quite small, with a volume of around 100 L. However, if the system needs to be able to start quickly in response to certain incidents, the expansion tank should be replaced by a load reservoir with the same capacity as the circuit. This reservoir may be heated by a vapor coil, or by the simple addition of live steam. The reservoir should not be considered as a replacement for the heating exchanger in the circuit, which is still required for use in the context of normal operations.

The expansion tank or reservoir is placed at a height such that sufficient net aspiration height is available for the pumps.

The flow through each tracer is adjusted by a manual valve located near the return collector. Contact thermometers and flowmeters (on collectors) are used for this purpose.

14

Combustion and Sulfur Dew Point

14.1. Characteristics of combustion

14.1.1. *Combustion air*

Dry air includes 20.9% of oxygen by volume. The remaining 79.1%, made up of nitrogen and around 1% of rare gases (mostly argon, krypton and xenon), is inert in terms of combustion.

Atmospheric air, however, is always humid to some degree, and may be characterized by its relative humidity ε at 20°C, corresponding to a partial pressure of water vapor equal to:

$$P_{H_2O} = \varepsilon \times 0.0238 \qquad \text{(atm)}$$

The composition by volume of air, considered at 1 atm abs, is therefore (by percentage):

Oxygen	$Y_0 = 20.9\left(1 - P_{H_2O}\right)$
Inert components	$Y_I = 79.1\left(1 - P_{H_2O}\right)$
Water vapor	$Y_V = P_{H_2O} \times 100$
Total	100

14.1.2. *Fuel*

Natural gas is desulfurized before use, although this is not always the case for coal and fuel oil used in ovens and boilers. The composition of solid or liquid combustibles is given by mass, often in the following form:

Hydrogen/carbon ratio = X

Sulfur in total dry product = S

Humidity, where applicable (carbon or fuel oil burner with added water vapor) in the total wet (coal) or dry (fuel oil).

The composition of a dry combustible, by percentage, is:

$$X_H = \left[100\,X/(1+X)\right](1-S)$$

$$H_C = \left[100/(1+X)\right](1-S)$$

$$X_S = S$$

The mass of water which accompanies 1 kg of *dry* fuel is:

provided directly by fuel oil: M_{H_2O}

for coal, with humidity h (weight fraction): $M_{H_2O} = \dfrac{h}{100-h}$

Fuel oil contains less than 1% of the (oxygen + nitrogen) couple, and negligible proportions of natural humidity and ash.

Coal may include 7–10% ash and a similar quantity of water. The proportion of oxygen + nitrogen may be as high as 3%. Strictly speaking, the (oxygen + nitrogen) content in the resulting smoke should be taken into account. The calculations given below are for (fictional) de-ashed coal.

14.1.3. *Stoichiometric air requirement*

The stoichiometric air requirement is the volume of air required for stoichiometric combustion of 1 kg of fuel.

Given the atomic masses of sulfur, carbon and hydrogen (32, 12 and 1, respectively), and given that the volume of 1 kmol of perfect gas at 0°C and under a pressure of 1 atm is 22.42 Nm^3, the stoichiometric air requirement for 1 kg of fuel is, in Nm^3:

$$V_a = \frac{22.42}{Y_0}\left[\frac{X_C}{12}+\frac{X_H}{4}+\frac{X_S}{32}\right]$$

14.1.4. *Smoke generation and smoke composition by volume*

Let λ be the excess air ($0 < \lambda > 1$). Using this value, we obtain the following volumes for 1 kg of fuel:

Inert components: $V_a(1+\lambda)Y_I/100$

Water vapor: $V_a(1+\lambda)\times\frac{Y_V}{100}+\frac{X_H}{200}\times 22.42+M_{H_2O}\times\frac{22.42}{1800}$

Oxygen: $\lambda V_a\times\frac{Y_0}{100}$

Carbon dioxide: $\frac{X_C}{12}\times 22.42$

Sulfur dioxide: $\frac{X_S}{32}\times 22.42$

The composition by volume is then very simple to deduce.

14.2. SO_3 content and dew point

14.2.1. *Calculation of the partial pressure of sulfur trioxide*

SO_2 is converted to SO_3 following the equation:

$$SO_2+\frac{1}{2}O_2 \rightleftarrows SO_3$$

The law of mass action for this balanced reaction is written:

$$P_{SO_3} = KP_{SO_2} \times P_{O_2}{}^{1/2}$$

The equilibrium constant K is given by the relationship:

$$\log_{10} K = \frac{4956}{T} - 4.678$$

The equilibrium establishment velocity becomes negligible at temperatures of below 1,200°C. Calculating the constant K at this temperature, we obtain an upper limit for K and, consequently, for P_{SO_3}, i.e. the sulfur dew point. Hence, if T = 1,473 K, then:

$$\log_{10} K = \frac{4956}{1473} - 4.678$$

$$K = 0.0486$$

The partial pressures of SO_2 and O_2 before conversion shall be noted:

$$P_{SO_2}^{(0)} \quad \text{and} \quad P_{O_2}^{(0)}$$

Furthermore, we shall consider that:

$$P_{SO_3} = RP_{SO_2}$$

Hence:

$$P_{O_2} = P_{O_2}^{(0)} - 2R\,P_{SO_2}^{(0)} \qquad P_{SO_2} = P_{SO_2}^{(0)} - R\,P_{SO_2}^{(0)} = (1-R)P_{SO_2}^{(0)}$$

After squaring, the law of mass action is written as:

$$R^2 = K^2(1-R)^2\left[P_{O_2}^{(0)} - 2R\,P_{SO_2}^{(0)}\right]$$

If, as a first approximation, R is considered negligible in relation to 1, then:

$$R^2 + 2\left(K^2 P_{SO_2}\right)R - K^2 P_{O_2} = 0$$

$$R^{(1)} = -K^2 P_{SO_2} + \sqrt{K^4 P_{SO_2}^2 + K^2 P_{O_2}}$$

We take:

$$K^{(1)} = \left(1 - R^{(1)}\right)K$$

Hence:

$$R^{(2)} = -\left[K^{(1)}\right]^2 P_{SO_2} + \sqrt{\left[K^{(1)}\right]^4 P_{SO_2}^2 + \left[K^{(1)}\right]^2 P_{O_2}}$$

and

$$P_{SO_3} = R^{(2)} P_{SO_2}^{(0)}$$

14.2.2. *Calculating the sulfur dew point temperature*

Choumer's approximated formula [CHO 71]:

$$t_R = 257 + 12.2 \text{ Ln } P_{H_2O} + 7.8 \text{ Ln } P_{SO_3}$$

t_R: degrees Celsius

P_{H_2O} and P_{SO_3}: atm

EXAMPLE.–

Combustion of fuel oil with a 2% sulfur content and 11.6% hydrogen content by mass, with 12% excess air. The fuel oil is presumed to be dry, and combustion is carried out without adding water vapor.

The relative humidity of the air is 60% at 20°C.

The composition of the fuel oil is:

X_H	=	$11.6 \times (1 - 0.02)$	=	11.37
H_C	=	$88.4 \times (1 - 0.02)$	=	86.63
X_S	=			2
			Total	100.00

$$P_{H_2O} = 0.6 \times 0.0238 = 0.0143 \text{ atm}$$

Giving the composition of the air:

Y_0	=	20.9 (1 – 0.0143)	=	20.6011
Y_I	=	79.1 (1 – 0.0143)	=	77.9689
Y_V	=		=	1.4300
			Total	100.00

Stoichiometric air requirement:

$$V_a = \frac{22.42}{20.6011}\left(\frac{86.63}{12} + \frac{11.37}{4} + \frac{2}{32}\right)$$

$$V_a = 11.0180 \text{ Nm}^3.\text{kg}^{-1}$$

Giving the smoke development power:

Inert components of exhaust smoke:	$11.018(1+0.12)\times 0.779689$	=	9.6756
Water vapor:	$11.018\times 1.12\times 0.0143+\frac{0.1137}{2}\times 22.42$	=	1.451
Oxygen:	$11.018\times 0.12\times 0.206011$	=	0.2724
Carbon dioxide:	$0.8663\times 22.42/12$	=	1.6185
Sulfur dioxide:	$0.02\times 22.42/32$	=	0.0140
	Total		13.0315

This gives the following partial pressures (in atm):

$$P_{H_2O} = 1.451/13.0315 = 0.11134 \quad \text{atm}$$

$$P_{O_2} = 0.2724/13.0315 = 0.02090 \quad \text{atm}$$

$$P_{SO_2} = 0.014/13.0315 = 0.00107 \quad \text{atm}$$

$$R^{(1)} = -(0.0486)^2 \times 0.00107 + \sqrt{(0.0486)^4 \times 0.00107^2 + (0.0486)^2 \times 0.0209}$$

$$R^{(1)} = 0.007028$$

$$K^{(1)} = 0.0486(1-0.007028) = 0.04826$$

$$R^{(2)} = -0.00000249 + 0.006977$$

$$R^{(2)} = 0.006979 \qquad P_{SO_3} = 0.00107\times R^{(2)} = 7.47\times 10^{-6}\,\text{atm.}$$

Following Choumer's formula [CHO 71]:

$$t_R = 257 + 12.2\ \text{Ln}0.11134 + 7.8\ \text{Ln}7.47 \times 10^{-6}$$

$$t_R = 138°\text{C}$$

15

Heat Supply by Microwave or Infrared Radiation

15.1. Microwave heating (theory)

The purpose of this section is to provide a general overview. A more detailed presentation may be found in Thuéry [THU 80].

15.1.1. *Maxwell's equations*

There are four of these equations:

$$\text{rot}\vec{H} = \vec{J} + \frac{\overrightarrow{\partial D}}{\partial \tau} = \sigma\vec{E} + \frac{\overrightarrow{\partial D}}{\partial \tau} \qquad (\text{Ampère's law}) \qquad [15.1]$$

$$\text{rot}\,\vec{E} = -\frac{\overrightarrow{\partial B}}{\partial \tau} \qquad (\text{Faraday's law}) \qquad [15.2]$$

$$\text{div}\,\vec{D} = \rho \qquad [15.3]$$

$$\text{div}\,\vec{B} = 0 \qquad [15.4]$$

Equations [15.2] and [15.4] are known as Maxwell's mechanical equations.

The magnetic induction $\vec{B}$ is defined by:

$$\vec{B} = \mu\vec{H}$$

with:

$\vec{H}$: magnetic field

μ: magnetic permeability (in a vacuum: $\mu = \mu_0 = 4\,\pi\times 10^{-7}$ S.I.)

The electrical induction $\vec{D}$ is defined by:

$$\vec{D} = \varepsilon\vec{E}$$

with:

$\vec{E}$: electric field

ε: electrical permittivity (in a vacuum: $\varepsilon = \varepsilon_0 = 1/(36\pi\times 10^9)$ S.I.)

NOTE.–

Some of the notations given above, while useful, have fallen out of general usage. The main elements which are still used are:

– the vector $\vec{B}$, now known as the magnetic field rather than the magnetic induction. Field H was measured in A m^{-1}, and B is measured in Tesla;

– vector $\vec{E}$, still known as the electrical field. Field E is measured in V m^{-1} and the electrical induction was measured in C.m^2 (Coulombs.m^2).

15.1.2. *Energy balance of electromagnetic radiation*

By combing Ampère and Faraday's laws, we obtain:

$$\vec{H}\ \text{rot}\ \vec{E} - \vec{E}\ \text{rot}\ \vec{H} = -\vec{E}\varepsilon\frac{\partial\vec{E}}{\partial\tau} - \vec{H}\mu\frac{\partial\vec{H}}{\partial\tau} - \vec{J}.\vec{E}$$

The following remarkable identity can be applied to the first member of this equation:

$$\operatorname{div}\left(\vec{A} \wedge \vec{B}\right) = \vec{B}.\operatorname{rot} \vec{A} - \vec{A} \operatorname{rot} \vec{B}$$

$$\operatorname{div}\left(\vec{E} \wedge \vec{H}\right) = -\left[\varepsilon\vec{E}\frac{\partial\vec{E}}{\partial\tau} + \mu\vec{H}\frac{\partial\vec{H}}{\partial\tau}\right] - \vec{J}.\vec{E}$$

Integrating the result over the volume of the body in question, applying Ostrogradski's theorem to the first member:

$$\int_S \left(\vec{E} \wedge \vec{H}\right) dS = -\int_V \left[\varepsilon\vec{E}\frac{\partial\vec{E}}{\partial\tau} + \mu\vec{H}\frac{\partial\vec{H}}{\partial\tau}\right] dV - \int_V \vec{J}.\vec{E}.dV$$

Take:

$$\vec{P} = \vec{E} \wedge \vec{H}$$

$\vec{P}$ is the *Poynting vector* in this vector product: $\vec{E}$ comes before $\vec{H}$.

Additionally, we take:

$$u = \frac{1}{2}\left(\varepsilon\vec{E}.\vec{E} + \mu\vec{H}.\vec{H}\right) = \frac{1}{2}\left(\varepsilon_o E^2 + \frac{B^2}{\mu}\right)$$

u is the volume density of electromagnetic energy present in the body in question.

The balance given above may therefore be written:

$$-\int_S \vec{P} dS = \int_V \frac{du}{d\tau} dV + \int_V \vec{J}.\vec{E} dV$$

The first member is the inverse of the Poynting vector flow through the surface S of the product. The first term of the second member represents the accumulation of electromagnetic energy, and the second term represents the heat dissipation of the electromagnetic energy.

15.1.3. *Electric field propagation equation*

The following identity may be applied to the electric field:

$$\text{rot rot } \vec{E} = \text{grad div } \vec{E} - \Delta\vec{E}$$

Moreover:

$$\text{div } \vec{E} = \frac{\rho}{\varepsilon}$$

$$\text{curl } \vec{E} = -\frac{\overrightarrow{\partial B}}{\partial \tau} = -\mu \frac{\overrightarrow{\partial H}}{\partial \tau}$$

Hence:

$$-\mu \frac{\partial}{\partial \tau} \text{curl } \vec{H} = \overrightarrow{\text{grad}} \frac{\rho}{\varepsilon} - \overrightarrow{\Delta E}$$

But:

$$\text{curl } \vec{H} = \vec{J} + \frac{\overrightarrow{\partial D}}{\partial \tau} = \sigma\vec{E} + \varepsilon \frac{\partial \vec{E}}{\partial \tau}$$

Thus, in the absence of free electrical charges:

$$\mu\, \varepsilon \frac{\partial^2 \vec{E}}{\partial \tau^2} + \mu\, \sigma \frac{\partial \vec{E}}{\partial \tau} - \Delta\vec{E} = 0 \qquad [15.5]$$

Symmetrically, the equation obtained is the same as that for magnetic field propagation.

Now, consider propagation in a vacuum.

Let 0z be the axis along which propagation occurs. We shall consider that vectors $\vec{E}$, $\vec{E}$ and $\vec{H}$ are perpendicular to 0z (this will be verified later).

We therefore take:

$$E_z = 0 \quad E_x = E_{x0}\, e^{j(\omega\tau - \beta z)} \quad E_y = E_{y0}\, e^{j(\omega\tau - \beta z)}$$

$$H_z = 0 \quad H_x = H_{x0}\, e^{j(\omega\tau - \beta z)} \quad H_y = H_{y0}\, e^{j(\omega\tau - \beta z)}$$

A vacuum is an insulator for which $\sigma = 0$, and the propagation equation is written:

$$\frac{1}{\mu_o \varepsilon_o}\frac{\partial^2 E}{\partial z^2} - \frac{\partial E}{\partial \tau^2} = 0$$

Hence, inserting the expressions of E_x, E_y, H_z and H_y into this equation:

$$\frac{1}{\sqrt{\mu_0 \varepsilon_o}} = c = \frac{\omega}{\beta}$$

Faraday's Law is written:

$$\text{rot}\,\vec{E} = -\mu_o \frac{\partial \vec{H}}{\partial \tau}$$

Moreover:

$$\left(\text{curl}\,\vec{E}\right)_x = -\frac{\partial E_y}{\partial z} = j\beta E_y = -j\mu_o \omega H_x$$

$$E_{y0} = -\frac{\mu_o \omega}{\beta} H_{x0}$$

Similarly:

$$\left(\text{curl}\,\vec{E}\right)_y = \frac{\partial E_x}{\partial z} = -j\beta E_x = -j\mu_0 \omega H_y$$

$$E_{x0} = \omega \mu_0 H_{y0}$$

Hence:

$$\vec{E}.\vec{E} = E_{x0}^2 + E_{y0}^2 = \left(H_{x0}^2 + H_{y0}^2\right)\frac{\mu_0\omega^2}{\frac{\mu}{\beta}} = \vec{H}.\vec{H}\mu_0 c$$

$$\varepsilon_0\vec{E}.\vec{E} = \mu_0\vec{H}.\vec{H}$$

The bulk flow of electromagnetic energy in a vacuum is therefore:

$$u = \frac{1}{2}\left(\varepsilon_0\vec{E}.\vec{E} + \mu_o\vec{H}.\vec{H}\right) = \varepsilon_0\vec{E}.\vec{E}$$

Furthermore, $\vec{E}$ and $\vec{H}$ can be seen to be orthogonal, as:

$$\vec{E}.\vec{H} = E_{x0}H_{x0} + E_{y0}H_{y0} = 0$$

Similarly, we may verify that:

$$\left(\text{curl}\,\vec{E}\right)_z = 0 = H_{z0} \quad \text{and} \quad \left(\text{curl}\,\vec{H}\right)_z = 0 = E_{z0}$$

Vectors $\vec{E}$ and $\vec{H}$ are thus perpendicular to 0z, confirming our hypothesis.

15.1.4. *Penetration depth in a dielectric layer*

Equation [15.5] is written:

$$\frac{\partial^2 E}{\partial z^2} = \mu\varepsilon\frac{\partial^2 E}{\partial \tau^2} + \mu\sigma\frac{\partial E}{\partial \tau}$$

Take:

$$E = E_o e^{\gamma z + j\omega\tau}$$

After simplification, we obtain:

$$\gamma^2 = -\mu\varepsilon\omega^2 + \mu\sigma j\omega$$

Furthermore, taking

$$\gamma = \alpha + j\beta \quad \text{and} \quad \sigma = \omega\varepsilon tg\delta$$

We obtain

$$\alpha^2 - \beta^2 = -2j\alpha\beta + j\omega^2\mu\varepsilon tg\delta - \varepsilon\omega^2\mu$$

Making the real and imaginary parts of this equation equal to zero

$$\alpha^2 - \beta^2 + \varepsilon\mu\omega^2 = 0$$

$$2\alpha\beta - \omega^2\mu\varepsilon tg\delta = 0$$

Eliminating β

$$\beta = \frac{\omega^2\mu\varepsilon tg\delta}{2\alpha}$$

$$\alpha^2 - \frac{\left(\omega^2\mu\varepsilon tg\delta\right)^2}{4\alpha^2} + \varepsilon\mu\omega^2 = 0$$

Take:

$$\alpha^2 = A$$

$$A^2 + A\varepsilon\mu\omega^2 - \left(\frac{\omega^2\mu\varepsilon tg\delta}{2}\right)^2 = 0$$

$$A = \frac{-\varepsilon\mu\omega^2 + \sqrt{\left(\varepsilon\mu\omega^2\right)^2 + \left(\omega^2\mu\varepsilon tg\delta\right)^2}}{2}$$

$$\alpha^2 = A = \frac{\varepsilon\mu\omega^2}{2}\left[-1+\left(1+tg^2\delta\right)^{1/2}\right]^{1/2} > 0$$

Only the + sign is retained.

Similarly, we obtain:

$$\beta^2 = B = \frac{\varepsilon\mu\omega^2}{2}\left[1+\left(1+tg^2\delta\right)^{1/2}\right]$$

Finally:

$$\alpha = \omega\left[\frac{\mu\varepsilon}{2}\left(\left(1+tg^2\delta\right)^{1/2}-1\right)\right]^{1/2}$$

$$\beta = \omega\left[\frac{\mu\varepsilon}{2}\left(\left(1+tg^2\delta\right)^{1/2}+1\right)\right]^{1/2}$$

In a medium with a low conductivity σ, i.e. if $tg\delta$ is small in relation to 1:

$$\alpha = \frac{\omega\left(\mu\varepsilon\right)^{1/2}}{2}tg\delta \quad \text{and} \quad \beta = \omega\left(\mu\varepsilon\right)^{1/2}$$

Thus:

$$E = E_0 e^{-\alpha z + j(\omega\tau - \beta z)}$$

The field amplitude is divided by e = 2.718 when the distance traveled reaches $1/\alpha$. This distance is the penetration depth, d:

$$d = 1/\alpha$$

15.1.5. *Dielectric conductivity*

By definition, dielectrics have a very low level of conductivity; in other words, they contain very few moving charges.

However, under the influence of an alternating electrical field, the rotation of molecular dipoles slows due to the friction phenomena resulting from intermolecular connections. For this reason, dielectric losses may still be represented in the form of a dielectric "conductivity", σ_d:

$$\sigma_d = \omega\varepsilon \,\mathrm{tg}\delta \qquad [15.6]$$

The electrical permittivity ε of the dielectric is conventionally expressed as:

$$\varepsilon = \varepsilon_r \varepsilon_0 \qquad (\varepsilon_0 : \text{permittivity of vacuum})$$

For water, at a frequency of 3 GHz:

Temperature (°C)	ε_r	tgδ
1.5	80	0.31
95	52	0.047

For ice at 0°C:

$$\varepsilon_r = 3.2 \qquad \mathrm{tg}\delta = 0.0009$$

The relative permittivity ε_r was formerly known as the dielectric constant.

Polar chemical species such as water contain a permanent dipole, which produces heat as it rotates. However, induced polarization also plays a part. Finally, substances that have been dissolved and ionized in water move through the water, also creating heat through friction.

More precise data concerning water may be found in Nikwist *et al.* [NIK 75] and Hasted [HAS 79]. An extensive theoretical discussion of the behaviors of chemical species may be found in Thuéry [THU 89].

EXAMPLE 15.1.–

For a wet product:

$$\varepsilon = 70.\varepsilon_0 \qquad \mathrm{tg}\delta = 0.31$$

$$\mu = \mu_0 \qquad \omega = 2\pi \times 2.45\times10^9 \text{ rad.s}^{-1}$$

Furthermore:

$$\varepsilon_0 = 1/\left(36\pi\times10^9\right) \qquad \mu_0 = 4\pi\times10^{-7}$$

$$\alpha = 2\pi\times2.45\times10^9\left[\frac{4\pi\times10^{-7}\times70}{36\pi\times10^9\times2}\left[\sqrt{1+0.31^2}-1\right]\right]^{1/2}$$

$$\alpha = 65.7\ \text{m}^{-1}$$

Giving the penetration depth:

$$d = 1/65.7 = 0.015\ \text{m}$$

15.1.6. *Wave impedance*

By definition, wave impedance is:

$$\eta = \sqrt{\frac{E_x^2+E_y^2}{H_x^2+H_y^2}}$$

The Maxwell equations are written as:

$$E_x = E_{x0}e^{-\gamma z+j\omega\tau} \qquad E_y = E_{y0}e^{-\gamma z+\gamma\omega\tau}$$

$$H_x = H_{x0}e^{-\gamma z+j\omega\tau} \qquad H_y = H_{y0}e^{-\gamma z+\gamma\omega\tau}$$

$$\gamma H_y = \left(\sigma + j\omega\varepsilon\right)E_x \qquad -\gamma E_y = j\omega\mu H_x$$

$$-\gamma H_x = \left(\sigma + j\omega\varepsilon\right)E_y \qquad \gamma E_x = j\omega\mu H_y$$

But

$$\gamma^2 = j\omega\mu\left(\sigma + j\omega\varepsilon\right)$$

Hence:

$$\eta = \frac{\gamma}{\sigma + j\omega\varepsilon} = \sqrt{\frac{j\omega\mu}{\sigma + j\omega\varepsilon}}$$

If the conductivity is zero:

$$\eta = \sqrt{\frac{\mu}{\varepsilon}} \qquad [15.7]$$

For air:

$$\eta_0 = \sqrt{\frac{\mu_0}{\varepsilon_0}} = \left[4\pi\times10^{-7}\times36\pi\times10^{9}\right]^{1/2} = 377\Omega$$

For dielectrics, μ is very similar to μ_0, the value for a vacuum (and for air).

15.2. Microwave heating (practical)

15.2.1. *Energy consumption and heating time*

More than 90% of the electric power consumed by the generator is transformed into useful heat, in both domestic and correctly designed industrial ovens. The energy yield of a hot gas dryer is around 65%, meaning that the microwave option represents an improvement of (90/65) – 1 = approx. 40%.

Let M be the mass of product to heat from T_0 to T_1, C_p its specific heat capacity, and P the average power used by the generator. The heating time will be:

$$\tau_C = \frac{MC_P\left(T_1 - T_0\right)}{0.9\ P}$$

EXAMPLE 15.2.–

An 850 W oven is equipped with a generator, which *operates intermittently half of the time.* The average useful power, with a yield of 90%, is thus:

$$P_u = \frac{850}{2} \times 0.9 = 383 \text{ W}$$

A dish, comparable to a disc with a thickness of 3 cm and a diameter of 15 cm, must be heated from 20 to 85°C. The bulk density of the material is 600 kg m^{-3} and the heat capacity is 3,000 J $kg^{-1}°C^{-1}$.

The mass of the dish being heated is:

$$M = 0{,}03 \times (\pi / 4) \times (0.15)^2 \times 600 = 0.312 \text{ kg}$$

Giving a heating time:

$$\tau = \frac{0.312 \times 3000 \times (85 - 20)}{383} = 159 \text{ s}$$

$$\tau = 2 \text{ mn } 39 \text{ s}$$

15.2.2. *Heating power output: runaway*

The output power in terms of heat per unit of volume is:

$$\frac{P_d}{V} = \vec{E}.\vec{J} = \sigma_d \vec{E}.\vec{E}$$

Following equation [15.6], the dielectric conductivity is:

$$\sigma_d = \omega \varepsilon \text{tg}\delta$$

Thus:

$$\frac{P_d}{V} = \omega \varepsilon \text{tg}\delta \; \vec{E}.\vec{E}$$

However, the bulk density of energy from electromagnetic radiation in a vacuum is:

$$u_0 = \varepsilon_0 \; \vec{E}.\vec{E}$$

The power output per unit of volume is therefore:

$$\frac{P_d}{V} = \omega \frac{\varepsilon tg\delta}{\varepsilon_0} u_0 = \omega \varepsilon_r u_0$$

Moreover:

$$\varepsilon_r = \varepsilon_{r0} + \varepsilon_{r1}\left(T - T_0\right) + \varepsilon_{r2}\left(T - T_0\right)^2$$

The heat balance of an irradiated body of mass M, which loses thermal power q by convection is written as:

$$MC_p \frac{\partial T}{\partial \tau} + q = \omega \varepsilon_r P_0$$

Taking:

$$\theta = T - T_0$$

and:

$$n_0 = \frac{\varepsilon_{r0} P_0}{MC_p} \qquad n_1 = \frac{\varepsilon_{r1} P_0 - q}{MC_p} \qquad n_2 = \frac{\varepsilon_{r2} P_0}{Mc_p}$$

The heat balance equation is written as:

$$\int_0^T \frac{d\theta}{n_0 + n_1\theta + n_2\theta^2} = \tau$$

As the time τ tends to infinity, the temperature $T - T_0 = \theta$ must tend to a finite limit (root of the binomial). Therefore, to avoid runaway, the following condition must be respected:

$$n_1^2 - 4n_0 n_2 < 0$$

15.2.3. *Transmission coefficients between two dielectrics*

Consider two loss-free dielectrics, separated by a flat surface.

For a normal incidence, let i, r and t represent the incident, reflected and transmitted fields (from dielectric 1 toward dielectric 2):

$$E_i = \eta_1 H_i \qquad E_t = \eta_2 H_t \qquad E_r = -\eta_1 H_r$$

The continuity of fields (which are transversal in this case) is written as:

$$E_i + E_r = E_t \qquad \text{and} \qquad H_i + H_r = H_t$$

We may write:

$$H_i + H_r = \frac{1}{\eta_1}(E_i - E_r) = H_t = \frac{1}{\eta_2}(E_i + E_r)$$

Hence:

$$\frac{E_r}{E_i} \qquad \text{and} \qquad \frac{E_t}{E_i} = 1 + \frac{E_r}{E_i}$$

Finally:

$$\frac{E_t}{E_i} = \frac{2\eta_2}{\eta_1 + \eta_2} \qquad \text{and} \qquad \frac{H_t}{H_i} = \frac{2\eta_1}{\eta_1 + \eta_2} \qquad [15.8]$$

$$\frac{E_r}{E_i} = \frac{\eta_2 - \eta_1}{\eta_1 + \eta_2} \qquad \frac{H_r}{H_i} = \frac{\eta_1 - \eta_2}{\eta_1 + \eta_2}$$

The ratio of the energy flow densities is equal to the modulus ratio of the Poynting vectors:

$$\frac{\varphi_t}{\varphi_i} = \frac{E_t H_t}{E_i H_i} = \frac{4\eta_1\eta_2}{(\eta_1 + \eta_2)^2} \qquad [15.9]$$

$$\frac{\varphi_r}{\varphi_i} = \frac{E_r H_r}{E_i H_i} = \frac{-(\eta_1 - \eta_2)^2}{(\eta_1 + \eta_2)^2}$$

We can easily verify that:

$$\varphi_t = \varphi_r + \varphi_i$$

Thuéry [THU 89] discusses the case of an oblique incidence.

15.2.4. *Electric field in the oven and energy flow density*

The useful power is transmitted to the product in its entirety in the form of heat. Giving, in the center of the product:

$$E_p = \left[\frac{P_u}{V\ \omega\varepsilon\tan\delta} \right]^{1/2}$$

The transmission coefficient for the electric field from the chamber volume surrounding the product to the product is, according to equations [15.7] and [15.8]:

$$\frac{E_t}{E_i} = \frac{2\sqrt{\mu_0 / \varepsilon}}{\sqrt{\frac{\mu_0}{\varepsilon_0}} + \sqrt{\frac{\mu_0}{\varepsilon}}} = \frac{\frac{2}{\sqrt{\varepsilon_r}}}{1 + \frac{1}{\sqrt{\varepsilon_r}}}$$

EXAMPLE 15.3.–

$$\omega = 2\pi \times 2.45 \times 10^9 = 1.54 \times 10^{10}\,\text{rad.s}^{-1}$$

$$\tan\delta = 0,31 \qquad \varepsilon_r = 70$$

$$\varepsilon = \varepsilon_r \times \varepsilon_0 = \frac{70}{36\pi \times 10^9} = 0.619 \times 10^{-9}$$

$$P_u = 400\,\text{W} \qquad V = 5.2 \times 10^{-4}\,\text{m}^3$$

$$E_p = \left[\frac{400 \times 10^9}{5.2 \times 10^{-4} \times 1.54 \times 10^{10} \times 0.619 \times 0.31} \right]^{1/2} = 510 \text{ V.m}^{-1}$$

The transmission coefficient for the electric field between the product and the surrounding chamber volume is:

$$\eta_2 = \frac{E_t}{E_i} = \frac{2/\sqrt{70}}{1+1/\sqrt{70}} = 0.21 \quad \text{and} \quad \eta_1 = 1$$

The field in the chamber outside of the product is:

$$E_i = 510/0.21 = 2428 \text{ V.m}^{-1}$$

The energy transmission coefficient (see equation [15.9]) is:

$$\frac{\varphi_t}{\varphi_i} = \frac{4 \times 0.21 \times 1}{(0.21+1)^2} = 0.57$$

The energy flow density in the chamber surrounding the product is 1/0.57 = 1.75 times that encountered in the product.

15.2.5. *Practical arrangements and use*

Industrial electromagnetic microwaves operate in the 915 and 2,450 MHz bands, corresponding to c/N wavelengths from 0.01 to 0.1 m depending on the propagation environment.

These wavelengths are comparable to the size of objects being processed.

These waves are produced by a generator, which may be:

– a magnetron with circular symmetry, including a hollow cylindrical anode, containing, at its core, a heated tungsten cathode emitting electrons;

– a klystron, where the beam emitted by an electron gun travels through a series of resonance chambers.

The waves produced by the generator transfer their energy to the processed material through a system known as an applicator. This system is based on the waveguide principle, which will not be described in detail here.

Microwave technology offers a high energy yield and speed of operation for:

– thawing and/or cooking food products;

– freeze-drying and/or sterilizing these products;

– drying paper (ink, glue), leathers, coarse textile fibers, dyed fabrics, sugar-coated and granulated pharmaceutical products.

NOTE.–

When using microwave radiation to dry a bulk solid, the bulk solid must be agitated (fluidized, vibrated, etc.) continuously, otherwise the surface may burn before the inner part of the mass has dried.

15.3. Infrared drying

15.3.1. *Radiation production*

Infrared radiation may be categorized by wavelength:

Near-infrared (luminous)	0.75–3 μm
Mid-infrared (dark)	3–25 μm

Far-infrared (25–100 μm) is not used in drying food products, as they absorb radiation in the 2.5–10 μm interval.

Infrared drying techniques are mostly used for food products. Conversely, chemical products (in crystal form) are dried using a hot air current.

Three different sorts of infrared sources are used:

– a resistor in a quartz tube (λ # 1.2 μm);

– a bare corrugated resistor made with a special Kanthal alloy ($2.7 < \lambda < 4.3$ μm);

– resistors immersed in a powdered ceramic product (or magnesia) and protected by a cast-iron casing. These resistors are generally immersed in a liquid to heat it, and are known as immersion heaters. They are not used for drying.

The light flow density radiated onto the product has a value of around 4,500 W m^{-2}. This value is two to three times higher than the results obtained using hot air. Whilst it would be possible to obtain values in the region of 25,000 W m^{-2}, this is not desirable due to the temperature-sensitive nature of foodstuffs.

Radiation energy is characterized by its irradiance.

The total irradiance I is measured in W m^{-2}.

The spectral irradiance I_λ is measured in $W(m^2.\mu m)^{-1}$,

$$I_\lambda = \frac{\partial I}{\partial \lambda}$$

with λ the wavelength in question.

15.3.2. *Absorption of radiation*

The irradiance of radiation is weakened as it travels through an absorbent medium. We may write:

$$I = I_o \exp(-\sigma\mu)$$

μ: kg of matter traversed per m^2

σ: total extinction coefficient (m^2 kg^{-1})

The relationship I/I_o gives us the transmitted fraction t.

We define:

t: total transmissivity

r: total reflectivity

a: total absorptivity

and:

$$t + a + r = 1$$

If φ is the flow density produced by the source:

$$I_o = \varphi(1 - r)$$

r varies between 0.1 and 0.2.

Taking ρ_p as the bulk density of the wet product, the heat balance for a section of product is given by the relationship:

$$LE + \rho C_p \frac{\partial T}{\partial \tau} = -\lambda \frac{\partial^2 T}{\partial x^2} + \sigma \rho I_o \exp(\sigma \rho x) \qquad [15.10]$$

λ: thermal conductivity of the product (W/m°C)

L: latent heat of water: (J kg^{-1})

E: bulk density of vaporization (kg m^{-3} s^{-1})

x is the depth of the product in question, measured from the surface.

The second relationship takes account of the fact that the vaporizing pressure of the vaporized water is equal to the ambient pressure P_a.

$$P_a = a_e(X)\pi(T) \qquad [15.11]$$

$\pi(T)$: vaporizing pressure of pure water (Pa)

$a_e(X)$ is the activity of the water in the product, dependent on the local humidity X within the product. X is the humidity value for the whole.

Profiles T(x) and E(x) for any given instant may be obtained by numerical integration of equations [15.10] and [15.11]. These profiles enable us to deduce X(x), i.e. the variation in humidity X across a thickness x.

15.3.3. *Practical data*

For fruit and vegetables, total transmissivity is around 0.05 at a depth of 3 cm. These products have a water content of 70–85%. Often, as the water content diminishes (e.g. from 85 to 60%), the total transmissivity increases, and may even be doubled; however, this is not always the case.

The depth of attenuation may be defined as the depth at which total transmissivity reaches a value of 0.05. Infrared drying is used for products with a thickness of between 0.5 and 5 cm.

If the product thickness is greater than the depth of attenuation, then continuing irradiation to dry deeper zones is unwise, as the temperature of the outer layer will continue to increase, causing heat damage to the surface. Instead, it is better to stop the process periodically in order for the temperature in the product to even out. Our equations take account of this.

Evidently, it is best to irradiate a thick mass of solid from sources situated in multiple directions.

Depending on the product, the operation may take from 10 min up to several hours, and may be carried out in a heat chamber or a tunnel dryer.

The energy yield is defined as:

$$\frac{\text{Vaporized mass} \times \text{latent heat}}{\text{Electrical energy input}}$$

In the best cases, this yield may be close to 80%.

16

Freezing, Deep-freezing and Thawing

16.1. Introduction

16.1.1. *Product properties*

Schwartzberg [SCH 76] proposed a method to determine the thermal capacity of foodstuffs and of most biological systems for both freezing and thawing purposes.

Marin *et al.* [MAR 85] determined the conductivity, latent heat and density of frozen products, and the quantity of water effectively frozen in frozen foods.

Chen [CHE 85] considered the speed at which ice is formed, thermal capacity and enthalpy, for meat, fish and fruit juices. Measurements were carried out at temperatures between 20 and –40°C.

Equation [15] in Pham [PHA 87] provides a means of determining the bound water content of a frozen food product. Note that bound water will not freeze, even with a significant reduction in temperature.

NOTE.–

Certain researchers have made use of an idealized synthetic mixture in their work. This mixture (23% by mass of methylcellulose, 77% water) is known as Tylose, and is intended to replicate the behavior of food products.

16.2. Industrial freezing apparatus

Jeanne [JEA 83] provides a thorough overview of deep-freezing equipment. A similar presentation may be found in Rémy [RÉM 87].

16.2.1. *The deep-freezing principle*

The temperature of the product must reach a value less than or equal to −18°C.

In most cases, cooling should be as fast as possible in order to create small ice crystals; larger crystals will deform the product, altering its quality.

16.2.2. *Batch freezer cabinets*

Using this system, a product is placed on shelves and cooled using cold air.

In this case, the hardware is relatively cheap, but a considerable amount of manpower is required.

In some systems, the product is placed onto double-walled plates filled with a vaporizing coolant; in this case, the "air blast" technique is not required.

16.2.3. *Tunnel freezers*

These systems are suitable for operations taking place over several hours. The product is placed onto aluminum plates or trays, which are then blasted with cold air from above and below. Fruit should be arranged in a single layer to prevent them from sticking together.

The plates or trays are placed onto trolleys, which move slowly through a tunnel. Manpower is required to load and unload the trolleys.

16.2.4. *Continuous blast freezers*

In these systems, input and output are continuous. Air is blasted onto the product from the side.

Depending on the residency time, the transport mechanism may be:

– a conveyor belt (for times of up to 10 min);

– a spiral (tens of minutes);

– an automatic chain conveyor (for times over an hour).

16.2.5. *Contact freezing*

The product is placed into direct contact with a planar surface, which may be horizontal, vertical or even curved (like the surface of a drum).

The flat surface freezing process is generally carried out in batches, while drum mechanisms can be used as part of a continuous system. In this case, the product is frozen in one full turn of the drum, taking from 1 to 15 min. Production rates vary between 25 and 100 kg h^{-1} per square meter of external drum surface.

Heat exchange may occur via contact with a flat metallic strip, acting as a conveyor. This type of mechanism may be used for pasty liquids (such as liquid sulfur), but this situation does not involve freezing.

16.2.6. *Freezing with refrigerant spray*

In this operation, liquid nitrogen or liquid carbon dioxide, sprayed to produce dry ice, are used.

This technique may be used with two types of equipment:

– tunnels;

– cabinets.

16.2.7. *Deep freezing with refrigerant spray*

The system has a small footprint, and no weight is lost through evaporation (using air, between 4 and 5% is lost).

Liquid CO_2 is injected into the system, forming dry ice and gas in a rotating drum or on a conveyor belt.

Liquid nitrogen may be used instead of CO_2 (in this case, the heat transfer coefficient is around 80 W m^{-2}°C^{-1}).

16.2.8. *Continuous double band freezer*

These systems are used for paste-like products, sandwiched between two bands, cooled by spraying with propylene glycol.

16.2.9. *Fluidized or fixed bed freezers [MIC 85]*

Separated product (e.g. peas or strawberries) may be placed onto a perforated transport belt, through which cold air is blasted. This is known as a fixed bed freezer.

A fixed grid may also be used to inject air more quickly than in the case of a fixed bed freezer. Using this mechanism, the product is processed in a fluidized state, and travels along the grid like a fluid.

If the resting height of a fluidized bed is taken as equal to that of a fixed bed, more mechanical power is required to ventilate a fluidized bed than a fixed bed. As the heat transfer coefficients are similar in the two cases, the beds will be of the same length, and the productivity of the systems is very similar.

However, the fluidized bed system prevents product particles from sticking together. Reynoso and Calvelo [REY 85] provide a detailed comparison of these two configurations.

16.2.10. *Cooling units*

When designing a unit of this type:

1) A condenser cooling fluid is used (e.g. water at 20°C). From this, we deduce the operating temperature of the condenser and, subsequently, its pressure.

2) The cooling unit is used to produce air, brine or an ethylene glycol solution at a specific temperature. The boiling temperature of the coolant should be 2–3°C lower. To produce "iced" water (at 5°C, for example), the temperature of the wall should not be lower than 0°C. Given the boiling temperature of the liquid, we can calculate the pressure of the "evaporator" and the corresponding thermal power (i.e. the coolant flow).

3) We then select a compressor capable of increasing the pressure of the gas from that used in the evaporator to that used in the condenser. This corresponds to a certain mechanical power.

4) Note that the thermal power of the condenser is the sum of the mechanical pressure of the compressor and the thermal power of the evaporator.

5) For low temperatures (e.g. –40°C), two units should be used in sequence. The evaporator in the "hot" unit acts as a condenser in the "cold" unit.

16.2.11. *Coolant fluids*

These fluids circulate through cooling units following a predetermined cycle. Historically, chlorofluorocarbons (CFCs) were generally used. Ammonia is no longer used for hygiene reasons. Alkanes, such as ethane and propane, are used in the oil industry.

Liquid refrigerants should not be confused with coolant fluids, which simply transport cold from the cooling unit to the place of use, and are essentially aqueous solutions of mono- or polyethylene glycol, or calcium chloride.

The thermodynamic properties of liquid refrigerants can be readily obtained from the Institut du Froid in Paris, France.

16.2.12. *Thermal balance*

To calculate the power required from a cooling unit, we need to know the thermal characteristics of the product being frozen.

16.3. Freezing time (Planck's equation [PLA 41])

16.3.1. *Case of a plate*

The thermal power lost by the plate through convection is:

$$q = \alpha S(T_s - T_a)$$

S: surface of the plate (m^2)

T_s and T_a: surface temperature and ambient temperature (°C)

α: convection coefficient ($W\ m^{-2}°C^{-1}$)

Let x be the distance between the surface and the part of the product, which is not yet frozen. The power given above is transmitted according to Fourier's law:

$$q = S(T_m - T_s)\frac{\lambda}{x}$$

λ: thermal conductivity of the frozen part ($W\ m^{-1}°C^{-1}$)

T_m: average temperature of the freezing front (°C)

Eliminating T_s from the two equations:

$$q = \frac{S(T_m - T_a)}{\frac{1}{\alpha} + \frac{x}{\lambda}}$$

The heat lost increases the size of the frozen area:

$$q = SL\rho\frac{dx}{d\tau}$$

$L\rho$: specific freezing heat ($J\ m^{-3}$)

Eliminate q from these two equations, and integrate x, which varies from zero to D/2, where D is the total thickness of the plate, presumed to be cooled from both sides. The time τ_c required for freezing is:

$$\tau_c = \frac{\rho L}{T_m - T_a}\left[\frac{D}{2\alpha} + \frac{D^2}{8\lambda}\right]$$

16.3.2. *Case of a sphere*

1) Convection:

$$q = \alpha\pi D^2 (T_s - T_a)$$

2) Conduction:

$$q = -4\pi r^2 \lambda \frac{dT}{dr}$$

Integrating for values of r varying from D/2 to zero:

$$q = \frac{4\pi\lambda(T_s - T_m)}{\frac{1}{r} - \frac{2}{D}}$$

3) Eliminating T_s:

$$q = \frac{T_m - T_a}{\frac{1}{\pi}\left[\frac{1}{\alpha D^2} + \frac{1}{4\lambda r} - \frac{1}{2\lambda D}\right]}$$

4) Frozen front:

$$q = L\rho 4\pi\, r^2 \frac{dr}{d\tau}$$

5) Eliminating q from the two equations, and integrating:

$$\tau_c = \frac{\rho L}{T_m - T_a}\left[\frac{D}{6\alpha} + \frac{D^2}{24\lambda}\right]$$

16.3.3. *Generalization I*

We may write:

$$\tau_c = \frac{\rho L}{T_m - T_a}\left[\frac{PD}{\alpha} + \frac{RD^2}{\lambda}\right]$$

With:

	P	R
Plate	1/2	1/8
Cylinder	1/4	1/16
Sphere	1/6	1/24

D is the diameter of the sphere or cylinder, or the total thickness of the plate.

16.3.4. *Generalization II*

We may also write:

$$\tau_c = \frac{\rho L}{(T_m - T_a)}\frac{V}{S\alpha}\left[1 + \frac{Bi}{4}\right]$$

V: volume of the body (m^3)

S: surface subject to cooling (m^2)

Bi: Biot number:

$$Bi = \frac{\alpha D}{\lambda}$$

16.3.5. *Freezing and thawing times (numerical calculation)*

Calculations have been carried out for the following simple forms:

– a plate of thickness D;

– a cylinder with a circular cross-section of diameter D;

– a sphere of diameter D.

Cleland and Earle [CLE 87] favored the finite element method. Equations for this method may be found in Cleland and Earle [CLE 84, Table 2].

Cleland *et al.* [CLE 86] studied thawing times, proposing four simple and precise empirical formulas, involving:

– the Fourier number, Fo;

– the Stéfan number, Ste;

– the Planck number, Pk;

– the Biot number, Bi.

16.4. Freezing time: practical method

16.4.1. *Introductory note*

The methods described above are only valid for simple geometric forms [CLE 87a]. The authors then generalized these results for more complex forms [CLE 87b].

16.4.2. *Definitions of characteristic dimensions*

1) Flat cylinder ("camembert" shape):

The key dimension D is the thickness. With ϕ the diameter of the cylinder, we take:

$$\beta_1 = \beta_2 = \phi/D$$

2) Cylinder of finite length, circular cross-section:

D is the diameter. Take:

$$\beta_1 = D/D = 1$$

If H is the length of the cylinder:

$$\beta_2 = H/D$$

3) Prism of finite length, rectangular cross-section:

D is the shortest side of the cross-section.

L is the longest side of the cross-section:

$$\beta_1 = L/D$$

If H is the length of the prism:

$$\beta_2 = H/D$$

4) Parallelepiped:

D corresponds to the shortest edge.

ℓ corresponds to the intermediate side.

$$\beta_1 = l/D$$

L corresponds to the longest edge.

$$\beta_2 = L/D$$

5) Prism of infinite length, rectangular cross-section:

D is the shortest side.

L is the longest side of the cross-section:

$$\beta_1 = L/D$$

$$\beta_2 = \infty$$

6) To summarize:

β_2 is the relationship between the longest dimension and the shortest dimension.

β_1 is the relationship between the intermediate dimension and the shortest dimension.

16.4.3. *Cleland et al.'s method (1987)*

Firstly, we calculate:

$$\tau_o = \frac{\rho L}{T_m - T_a}\left[\frac{D}{2\alpha} + \frac{D^2}{8\lambda}\right]$$

This reference time corresponds to a plate that is cooled from both sides.

The freezing time is then obtained using Cleland *et al.*'s equation [4] [CLE 87a].

$$\tau_c = \tau_o \frac{2V}{SD}\left[\frac{1 + \frac{\text{Bi LCM}}{4D}}{1 + \frac{\text{Bi}}{4}}\right]$$

V: volume of the solid (m^3)

S: surface subject to cooling (m^2)

MCP: mean conduction pathway (m)

$$\frac{MCP}{D} = 1 + (2 - G_1)M_1 + G_3M_2 + M_3$$

The MCP is given by equation [9] in the cited work.

The geometric parameters G_i are given in the following table (taken from Table 2 in the cited work):

	G_1	G_2	G_3
Plate	1	0	0
Infinite cylinder	2	0	0
Sphere	3	0	0
Flat cylinder	1	2	0
Short cylinder	2	0	1
Infinite prism	1	1	0
Parallelepiped	1	1	1

$$M_1 = \frac{2X_1G_2}{\beta_1^{1.92}} \times 0.63\mathrm{Bi}^{0.39} \times \left(1.34\ \beta_1^{1.43} - 1\right)$$

$$M_2 = \frac{2X_2G_3}{\beta_2^{1.92}} \times 0.63\mathrm{Bi}^{0.39} \times \left(1.51\beta_2^{1.22} - 1\right)$$

$$M_3 = \left[1 - \frac{\dfrac{2X_1G_2}{\beta_1^{1.92}} + \dfrac{2X_2G_3}{\beta_2^{1.92}}}{G_2G_3 + 1}\right]\left[\frac{G_1 + \dfrac{G_2}{\beta_1} + \dfrac{G_3}{\beta_2}}{G_1 + \dfrac{0.73G_2}{\beta_1^{2.5}} + \dfrac{0.50G_3}{\beta_2^{3.69}}} - 1\right]$$

The weighting coefficients are:

$$X_j = \frac{\dfrac{2}{\beta_j^{1.92}}}{\mathrm{Bi}^{1,34} + \dfrac{2}{\beta_j^{1.92}}}$$

In reality, Cleland *et al.* [CLE 87b, p. 240] give multiple expressions of X. These coefficients are used to calculate M_1, M_2 and M_3.

16.4.4. *Average freezing temperature*

Experimentally, we know that the frozen mass fraction of a food product increases as the temperature decreases. To carry out the calculations shown above, we need an average temperature T_m lower than the initial freezing temperature and higher than the ambient temperature T_a.

Pham states that this temperature T_m is dependent on the final temperature of the product T_f and on the ambient temperature T_a.

$$T_m = 1.8 + 0.263 T_f + 0.105 T_a$$

[PHA 86, equation [3]].

These temperatures are expressed in°C.

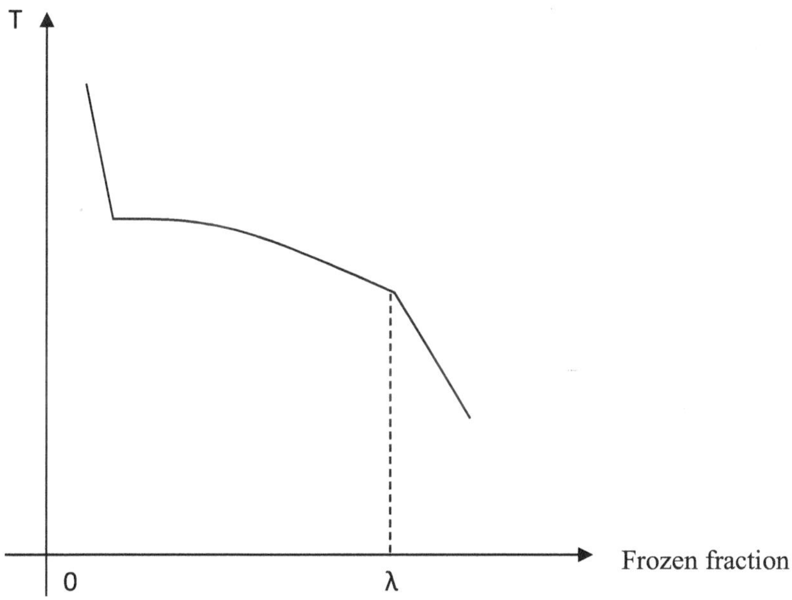

Figure 16.1. *Relationship between temperature and frozen fraction of product*

16.4.5. *Precooling correction*

With C_c and C_n, the thermal capacities by volume of the solid frozen product and the non-frozen product, we may write:

– energy density of precooling:

$$\Delta H_p = C_n \left(T_i - T_m\right)$$

T_i is the initial temperature of the product.

– freezing energy:

$$\Delta H_c = \rho L + C_c (T_m - T_f)$$

– mean temperature difference for precooling:

$$\Delta T_p = \frac{(T_i + T_m)}{2} - T_a$$

– mean temperature difference during freezing:

$$\Delta T_c = T_m - T_a$$

Pham [PHA 86] recommends replacing:

$$\text{factor} \frac{\rho L}{T_m - T_a} \quad \text{by} \quad \frac{\Delta H_p}{\Delta T_p} + \frac{\Delta H_c}{\Delta T_c}$$

EXAMPLE.–

Calculate the freezing time for beef with a starting temperature of 20°C descending to –18°C, in the form of parallelepipedic blocks of dimensions:

$$0.1 \text{ m} \times 1 \text{ m} \times 1 \text{ m}$$

The cooling air is at a temperature of –30°C.

$\Delta = 1.6 \text{ W.m}^{-1}.°\text{C}^{-1}$ $\qquad C_n = 3.22 \times 10^6 \text{ J m}^{-3}.°\text{C}^{-1}$

$\rho L = 2.55 \times 10^8 \text{ J m}^{-3}$ $\qquad C_c = 1.67 \times 10^6 \text{ J m}^{-3}.°\text{C}^{-1}$

$\alpha = 15 \text{ W m}^{-2}.°\text{C}^{-1}$

We calculate:

$$\text{Bi} = \frac{15 \times 0.1}{1.6} = 0.94$$

$$\beta_1 = \beta_2 = 10$$

$$X_1 = X_2 = \frac{\frac{2}{10^{1.92}}}{0.94^{1.34} + \frac{2}{10^{1.92}}} = 0.0254$$

$$G_1 = G_2 = G_3 = 1$$

$$M_1 = \frac{2 \times 0.0254}{10^{1.92}} \times 0.63 \times 0.94^{0.39}\left(1.34 \times 10^{1.43} - 1\right)$$

$$M_1 = 0.013$$

$$M_2 = \frac{2 \times 0.0254}{10^{1.92}} \times 0.63 \times 0.94^{0.39}\left(1.51 \times 10^{1.22} - 1\right)$$

$$M_2 = 0.009$$

$$M_3 = \left[1 - \frac{\frac{4 \times 0.0254}{10^{1.92}}}{2}\right]\left[\frac{1+0.2}{1+\frac{0.73}{10^{2.5}}+\frac{0.5}{10^{3.69}}} - 1\right]$$

$$M_3 = 0.2$$

According to equation [3] in the cited work:

$$\frac{MCP}{D} = 1 + 0.013 + 0.009 + 0.2 = 1.22$$

$$S = 2 \times (1 \times 1) + 0.1 \times (4 \times 1) = 2.4 \text{ m}^2$$

$$T_m = 1.8 + 0.263(-18) + 0.105(-30) = -6.1°C$$

Precooling energy:

$$\Delta H_p = 3.22 \times 10^6 \left[20 - (-6.1)\right] = 8.4 \times 10^7 \text{ J.m}^{-3}$$

Freezing energy:

$$\Delta H_c = 2.55\times10^8 + 1.67\times10^6\left\{-6{,}1-(-18)\right\} = 2.75\times10^8\,\text{J.m}^{-3}$$

Mean temperature difference for precooling:

$$\Delta T_p = \frac{1}{2}\left[20+(-6.1)\right]-(-30) = 37°\text{C}$$

Mean temperature difference for freezing:

$$\Delta T_c = -6.1-(-30) = 23.9°\text{C}$$

$$\frac{\Delta H_p}{\Delta T_p}+\frac{\Delta H_c}{\Delta T_c} = \frac{8.4\times10^7}{37}+\frac{2.75\times10^8}{23.9} = 1.38\times10^7\,\text{J.m}^{-3}.°\text{C}^{-1}$$

$$\tau_o = 1.38\times10^7\left[\frac{0.1}{2\times15}+\frac{0.01}{8\times1.55}\right]$$

$$\tau_o = 56670\text{ s}$$

$$\tau_c = 56670\times\frac{2\times0.1}{2.4\times0.1}\left[\frac{1+\dfrac{0.94\times1.22}{4}}{1+\dfrac{0.94}{4}}\right]$$

$$\tau_c = 49213\text{ s} = 13\text{ h}40\text{ min}$$

NOTE.–

Castaigne and Lacroix [CAS 85] developed a semiempirical method for calculating freezing time. This method makes a clear distinction between:

– precooling (before freezing);

– freezing "proper";

– overcooling (after freezing).

The time taken for the freezing element of the process is obtained by a simple use of Planck's equation [PLA 41].

NOTE.–

Certain authors consider temperature T_m to be the temperature at the center of the thermal capacity, which, for a homogeneous solid, corresponds to the center of mass (i.e. the barycenter).

16.5. Thawing

16.5.1. *Microwave thawing [ROS 88]*

The accepted frequency in France is 2,450 MHz.

The device is made up of wave-emitting tubes (magnetrons) with a power varying from 25 to 100 kW.

A waveguide (metallic cavity) is used to direct waves toward the substance being heated.

A wave stirrer is a fan with blades which reflect the waves.

This process is carried out for a short period of time, reducing exudation losses.

As soon as liquid water appears, it is more likely than the product to absorb the waves, which may lead to *surface overheating*. Products should therefore undergo a preheating process *before* thawing.

A phased process may also be used, i.e. on-off-on, etc. Another option is to introduce air circulating at –30°C or to use nitrogen spray in parallel with the microwave heating process.

The use of microwave technology is highly efficient in terms of bacteria limitation.

16.5.2. *External heating*

– In air, either at 20°C or at 4°C (in a cold room);

– Using cold water or water at 30°C, by spraying or immersion;

– Using vacuum steam at temperatures of between 10 and 30°C.

16.5.3. *Contact heating*

In this case, the product is brought into contact with a flat surface.

16.5.4. *Exudation losses*

Rosset and Roussel-Ciquart [ROS 83] provide an overview of exudation losses for different foodstuffs.

16.5.5. *Calculating thawing time*

Abdalla and Singh [ABD 85] use the finite element method for a simple geometric shape (axisymmetric cylinder). The authors considered the temperature of the geometric center of the shape (i.e. of the axis).

Cleland *et al.* [CLE 86] provide an evaluation of four prediction methods for simple-form solids. All four methods were found to produce results comparable to those obtained using numerical methods (finite elements and finite differences), all within 10% of experimental results.

Cleland *et al.* [CLE 87] list the precautions to take when using both of these numerical methods, in the context of both freezing and thawing. The finite element method is harder to use than the finite difference method.

17

Freeze-drying

17.1. General points

17.1.1. *The freeze-drying principle*

This operation consists of freezing a foodstuff or pharmaceutical product, before treating the cold product under a high vacuum. Condensed water in the product is converted into vapor and removed by the vacuum pump, thus drying the product.

At the end of the initial freezing phase, water is present in two forms:

– pure ice, in the form of crystals;

– the remainder, in the form of nonsolid (and at best, paste-like) mixtures. This phase is a concentrated solution in which water acts as the solvent. For this reason, water in this solution is known as solvent water.

As water in the product enters the gaseous phase, the ice and water components behave differently:

– the solvent water vaporizes with an enthalpy of vaporization corresponding to the passage from the liquid to vapor phase, on the condition that the dissolving heat of the solid in the solvent water is zero;

– the ice is sublimed. The enthalpy of sublimation is the sum of the enthalpy of fusion and the enthalpy of vaporization.

Freeze-drying therefore requires a significant amount of heat. The process is made even more complex by the fact that, during the drying process, the ice present in the product must be prevented from melting (final vaporization of the remaining solvent water may be carried out at a higher temperature).

Flink and Karel [FLI 72] studied the retention of volatile organic compounds in freeze-dried products.

To and Flink [TO 78] considered the transition of freeze-dried products at a temperature T_c. The solid matrix was found to break down, collapse and finally contract, taking on a glassy appearance. The authors identified the relationship between the contraction temperature and the loss of flavor (loss of taste and odor).

17.1.2. *Heat supply*

Heat can be provided in one of three ways:

1) internal microwave heating;

2) contact heating: considering the product as a plate, one face is placed into contact with a "hot" metallic surface, and the water vapor produced in this way is evacuated via the other face;

3) infrared heating: again, considering the product as a plate, one face is thermally insulated while the other is subjected to radiation; vapor then escapes from this surface.

The heat flow density transferred from a hot, flat plate at a temperature T_C to a planar surface at a lower temperature T_0 is given by:

$$q = \frac{\sigma\left(T_C^4 - T_0^4\right)}{\frac{1}{\varepsilon_C} + \frac{1}{\varepsilon_0} - 1} \qquad \left(\mathrm{W\,m^{-2}}\right)$$

ε_C and ε_0 are the emissivities of the two surfaces, and have values of around 0.8.

σ is Stéfan's constant: 5.67×10^{-8} W m^{-2} K^{-4}.

17.1.3. *Freeze-drying: implementation*

In the pharmaceutical industry, certain products are freeze-dried before being vacuum-sealed into ampoules or vials. In this case, the product is freeze-dried in batches, spread across trays, which are then placed into a horizontal cylindrical container.

In the food industry, the product most commonly subject to freeze-drying is coffee, as the process makes it instantly soluble; however, the same process can be used for soluble tea products or powdered milk. Figure 17.1 shows a continuous system where a concentrated solution is sprayed into a cold container, under vacuum, at a temperature close to its freezing point. Droplets are frozen, before falling onto a conveyor, which carries the product under infrared panels.

A vacuum device (ejector) removes enough water vapor from the liquid droplets so that the cooling process following evaporation causes the product to freeze.

The product is retrieved using a sluice-type mechanism.

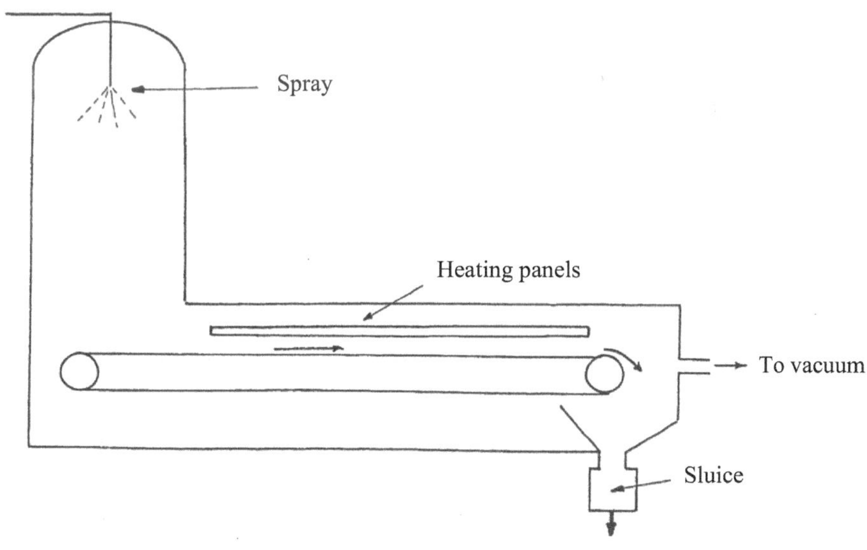

Figure 17.1. *Continuous freeze-drying*

17.2. Thermodynamics of freeze-drying

17.2.1. *Activity of solvent water in equilibrium with ice*

The ice and solvent water must be in equilibrium with the third phase, i.e. water vapor, at partial pressure P_V. Here, the three phases are at the same temperature T, and are in equilibrium if the vaporization pressures of the ice and solution are both equal to P_V:

$$P_V = \pi_G(T) = a_e \pi_L(T)$$

$\pi_G(T)$ and $\pi_L(T)$ are the vaporization pressures of the ice and liquid water (solvent) at temperature T.

Values of $\pi_G(T)$ are given in Table 17.1:

T (°C)	−40	−30	−20	−10	0
$\pi_G(T)$ (Pa)	12.83	38.33	134.59	260.84	610.9

Table 17.1. *Vaporizing pressure of ice*

Below 0°C, $\pi_L(T)$ is easier to calculate than to measure. To do this, consider the two different ways of passing from ice to solvent water at a shared temperature T (lower than 0°C):

1) the direct pathway;

2) a pathway passing through:

– ice at the triple point ($T_T = 0.01$°C),

– liquid water at the temperature of the triple point T_T,

– solvent water at T.

Assimilating the vaporization pressures to the fugacity values, at temperature T, we may write:

$$RTLn\frac{\pi_L}{\pi_G} = RTLn\frac{f_L}{f_G} = [\Delta G]_G^L = [\Delta H]_G^L - T[\Delta S]_G^L$$

Taking L_T as the specific fusion enthalpy of the ice at the triple point and C_L and C_G as the specific heat capacities of supercooled water and ice, using the second pathway, we obtain (with temperatures expressed in K):

$$\frac{[\Delta H]_G^L}{18} = C_G(T_T - T) + L_T - C_L(T_T - T)$$

$$= L_T - \Delta C(T_T - T) \qquad (\Delta C = C_L - C_G)$$

$$[\Delta S]_G^L / 18 = C_G \int_T^{T_T} \frac{dT}{T} + \frac{L_T}{T_T} + C_L \int_{T_T}^{T} \frac{dT}{T} = \frac{L_T}{T_T} - \Delta C \,\text{Ln}\frac{T_T}{T}$$

Thus:

$$RT \ln\frac{\pi_L}{\pi_G} = 18L_T\left(1 - \frac{T}{T_T}\right) - 18\Delta C(T_T - T) + 18\Delta C T \ln\frac{T_T}{T}$$

Finally, the activity of the solvent water is:

$$a_e = \frac{\pi_G}{\pi_L} = \exp\left[\frac{18L_T}{R}\left(\frac{1}{T_T} - \frac{1}{T}\right) + \frac{18\Delta C}{R}\left(\frac{T_T}{T} - 1 - \text{Ln}\frac{T_T}{T}\right)\right]$$

EXAMPLE 17.1 .–

$$\frac{\Delta C}{R} \# \frac{2200}{8314} = 0.26 \text{ kmol.kg}^{-1} \qquad \frac{L_T}{R} = \frac{335000}{8314} = 40 \text{ kmol.kg}^{-1}.\text{K}$$

$$(\text{for: } T = -20°\text{C} = 253 \text{ K})$$

$$\frac{\pi_G}{\pi_L} = \exp\left[18 \times 40\left(\frac{1}{273} - \frac{1}{253}\right) + 18 \times 0.26\left(\frac{273}{253} - 1 - \text{Ln}\frac{273}{253}\right)\right]$$

$$\frac{\pi_G}{\pi_L} = a_e = 0.82$$

NOTE (Properties of ice).–

thermal conductivity: 2.39 W m^{-1} K^{-1}

bulk density: 923 kg.m^{-3}

bulk thermal capacity: between 0°C and −10°C: 2,050 J kg^{-1}°C^{-1}

between −30°C and −40°C: 1,840 J kg^{-1}°C^{-1}

enthalpy melting: 335,000 J kg^{-1} (at 0.01°C).

17.2.2. *Empirical expression of water activity for prefreezing*

Schwartzberg [SCH 76] expressed the activity of water in foodstuffs in the following way:

$$a_e = \frac{(n_e - n_{el})}{(n_e - n_{el}) + n_s}$$

In this mixture:

$n_{e\ell}$: number of kmoles of free water

n_e: number of kmoles of bound water

n_s: number of kmoles of solid

In the same mixture, let:

– m_s be the mass of solid (excluding fats)

– m_e be the mass of water.

According to the *definition of the empirical coefficient* ℓ, the mass of bound water is:

$$m_{el} = \ell m_s$$

Consequently:

$$a_e = \frac{(m_e - \ell m_s)}{(m_e - \ell m_s) + \frac{18}{M_S} m_s} = \frac{m_e - \ell m_s}{m_e - \ell m_s + E m_s} = \frac{m_e - \ell m_s}{m_e + (E-1)\, m_s}$$

E is an empirical parameter.

Product	ℓ
Glucose	0.15 to 0.20
Saccharose	0.3
Meat and fish	0.24 to 0.27
Fruit and vegetables	0.18 to 0.25

Table 17.2. *Values of coefficient ℓ*

For meat and fish:

$$m_s = \frac{525}{1 - X_S} \qquad (X_S \text{ : fraction of solid by mass})$$

For fruit juices:

$$m_s = \frac{200}{1 + 1.25X_S}$$

a_e is thus expressed using two empirical parameters, ℓ and m_s. As the humidity of the dry product is $X = m_e/m_s$, we obtain:

$$a_e = \frac{X - \ell}{X + (E - \ell)} < 1$$

This expression is valid for values of X between ℓ and 19 (95% humidity).

In practice, ℓ is lower than E.

EXAMPLE 17.2 (Proportion of frozen water).–

A fruit is frozen at −20°C. The total humidity of the fruit is 90%, i.e. X = 90/10 = 9:

For this fruit:

$$\ell = 0.20 \qquad E = 0.25$$

$$a_e = \frac{X - 0.2}{X + (0.25 - 0.20)}$$

If $a_e = 0.82$ at –20°C, the corresponding value of X is:

$$X = \frac{E + 1(1 - a_e)}{1 - a_e} = \frac{0.25 + 0.20 \times 0.18}{0.18} = 1.53$$

The proportion of frozen water is:

$$1 - \left(\frac{1.53}{9}\right) = 0.83$$

Figure 17.2 shows the relationship between the freezing temperature and the proportion of frozen water:

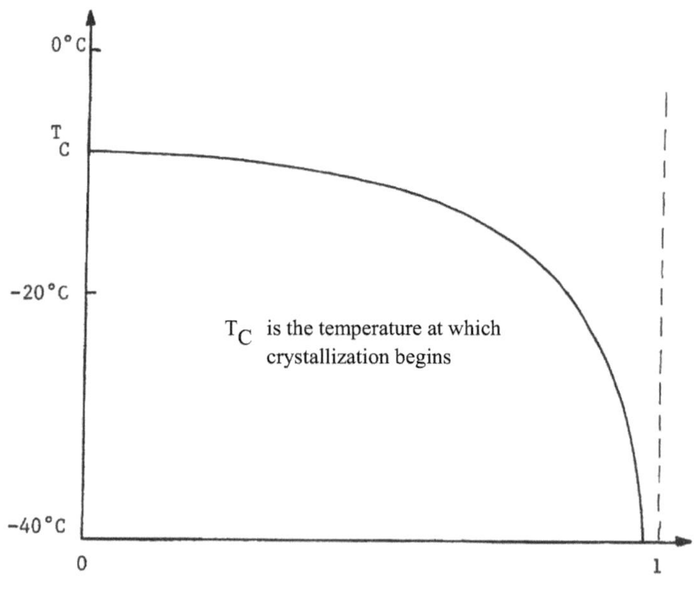

Figure 17.2. *Proportion of frozen water as a function of freezing temperature*

NOTE.–

As we have seen, for initial freezing, the vaporization pressure of solvent water may be expressed as:

$$a_e(X)\pi_L(T)$$

During the vacuum dehydration process used in freeze-drying, the humidity X will be reduced to well below the value observed after freezing. The analytical expression of a_e as a function of X for this phase of the operation will therefore be different to that obtained for the freezing process.

However, an empirical homographic function may still be used, and, for the drying phrase, we may write:

$$a_e(X) = \frac{A\,X + B}{C\,X + D}$$

The vaporization pressure of the solvent water is, as before, the product of the activity of the water and the vaporization pressure of pure water at temperature T:

$$P_V = a_e(X)\pi_L(T)$$

Note, however, that in extreme cases where X is close to zero, the analytical form of BET or other adsorption isotherms may be used. However, the physical hypotheses giving the analytical form of these isotherms are no longer valid, and so the expressions should be considered as empirical formulas.

17.2.3. *Expression of enthalpies for freeze-drying*

The specific thermal capacities of the different species and phases encountered are defined in Table 17.3.

Species or phase	Thermal capacity
Solid dry matter	C_S
Ice	C_{GL}
Solvent water	C_L
Water vapor	C_V
Air	C_A

Table 17.3. *Thermal capacities*

Taking 0°C as the origin, we obtain the following values for enthalpies (Table 17.4).

	Enthalpy
Solid dry matter	$C_S T$
Ice	$C_{GL} T$
Air	$C_A T$

Table 17.4. *Enthalpies*

The enthalpy of water vapor is obtained by subliming ice at 0°C and cooling the vapor to temperature T:

$$H_V = L_S + C_V T$$

L_S: enthalpy of sublimation at 0°C

Similarly, for solvent water:

$$H_L = L_F + C_L T$$

L_F: enthalpy of melting of ice at 0°C

The enthalpy of the frozen zone is obtained using the average of the component enthalpies (ε: fraction of the volume occupied by the gaseous phase).

$$H_C = \rho_S (1-\varepsilon)\left[C_S T + X(L_F + C_L T) + Y C_{GL} T\right]$$

The enthalpy of the de-iced zone is:

$$H_D = \rho_S(1-\varepsilon)\left[C_S T + X\left(L_F + C_L T\right)\right] + \frac{\varepsilon P_T}{R\,T}\left[\underset{\text{air}}{M_A y_A C_A t} + \underset{\text{vapor}}{M_V y_V\left(L_S + C_V T\right)}\right]$$

These two enthalpies are expressed in joules/m^3 and are used in writing the energy equation.

17.2.4. *Conclusion: the freeze-drying process*

The product is frozen at a temperature that defines:

– the fraction of the water which is transformed into ice;

– the remainder, in the form of nonsolid (and at best, paste-like) mixtures. This phase is a concentrated solution in which water acts as the solvent. The water in this solution is solvent water.

The lower the freezing temperature, the closer the fraction of water transformed into ice will be to 1.

Before the outset, the product may be porous, and air may be contained in these pores. Prior to sublimation, the product is at the freezing temperature T, may be porous, and contains water, mostly (but not entirely) in the form of ice.

However, we shall begin by considering the product as a nonporous frozen block.

Two hypotheses may be used when considering the behavior of the product:

1) Ice disappearing in a uniform manner through the whole thickness of the product.

2) A receding ice front.

Using the first hypothesis, for vapor to escape toward the free surface, the partial pressure P_V must decrease in the same way in order to prevent pressure loss via percolation through the product. However, P_V is equal to the vaporization pressure of the ice, which is an increasing function of temperature. Temperature T must therefore decrease approaching the free surface. T can only increase via radiation heating from the free surface, as shown in Figure 17.3.

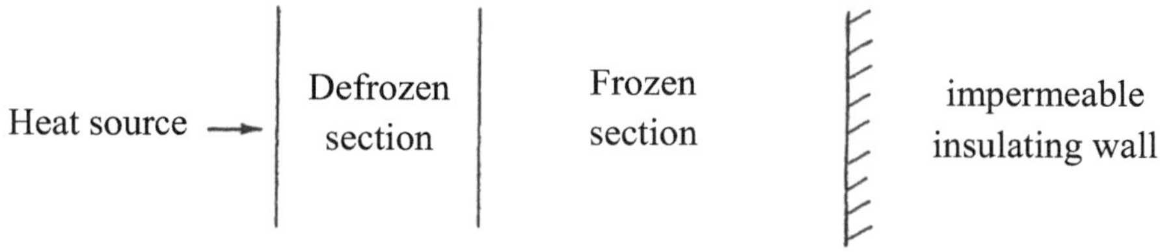

Figure 17.3. *Freeze-drying mechanism*

If, instead, heating is carried out by conduction through the insulating wall, the first hypothesis would be plausible. However, this situation will not be considered here, and we shall make use of the second hypothesis. A receding ice front separates the frozen zone from the ice-free zone, referred to as the "de-iced" (and not defrosted) zone.

The speed of sublimation is limited by the loss of pressure through the de-iced porous medium. An equilibrium between the heat input and vapor percolation must be attained.

Unlike situations which involve drying a porous medium wetted by a liquid, the solvent water (although not frozen) does not migrate, and is vaporized on the spot. This makes the problem considerably simpler.

17.3. Migration equations

17.3.1. *Free mean path*

In gas kinetics, the free mean path of molecules is:

$$\lambda = \frac{1}{\sqrt{2}} \frac{1}{\pi \delta^2} \frac{kT}{P}$$

P: pressure (Pa)

k: Boltzmann constant: $k = 1.38 \times 10^{-23} \text{J K}^{-1}$

δ is the collision diameter:

Air: 0.362 nm

Water vapor: 0.265 nm

EXAMPLE 17.3.–

P = 20 Pa (a normal pressure for freeze-drying)

T = –20°C = 253 K.

For water vapor:

$$\lambda = \frac{1}{\sqrt{2}} \times \frac{1}{\pi \times \left(0.265 \times 10^{-9}\right)^2} \times \frac{1.38 \times 10^{-23} \times 253}{20}$$

$$\lambda = 0.56 \times 10^{-3}\,\text{m}$$

This value of λ is considerably greater than the radius of the pores of a freeze-dried product. The situation is thus in the Knudsen regime.

17.3.2. *Flow through a tube in the Knudsen regime*

In a classic laminar regime, the gas velocity is:

$$\frac{q}{\pi R_0^2} = \frac{R_0^2}{8\mu}\frac{\Delta P}{L} = \frac{K_0}{\mu}\frac{\Delta P}{L}$$

R_0: radius of the pore (m)

In the Knudsen regime, the previous expression needs to be multiplied by the following factor:

$$F = \frac{8\lambda}{R_0}\sqrt{\frac{2}{\pi}}$$

i.e. multiplying F by the permeability K_0 of the pore (see [BRO 46]).

17.3.3. *Viscosity of a gas*

The mean velocity of the molecules in a gas is:

$$v^* = \left(\frac{8RT}{\pi M}\right)^{1/2} \qquad \left(m.s^{-1}\right)$$

R and M are the perfect gas constant and the molar mass of the gas, respectively.

The viscosity of the gas is obtained using:

$$\eta_G = \frac{1}{3}\rho_G v^* \lambda$$

ρ_G is the density of the gas.

17.3.4. *Migration equations*

For water vapor, the Maxwell-Stefan equation is written as:

$$-\nabla\mu_V - \bar{V}_V\ \nabla P_T = y_A \frac{RT}{D_{AV}}\left(\frac{J_V}{c_V} + \frac{J_A}{c_A}\right) + \frac{RT}{D_{VP}}\frac{J_V}{c_V}$$

A similar equation may be written for air. The following notation is used:

μ_V: chemical potential of water vapor (J $kmol^{-1}$)

$\bar{V}_V$: molar volume of water vapor (m^{-3} $kmol^{-1}$)

P_T: total pressure of the gaseous phase (Pa)

y_A: molar fraction of air in the gaseous phase

R: perfect gas constant: 8,314 (J $kmol^{-1}$ K^{-1})

T: absolute temperature (K)

D_{AV}: mutual diffusion coefficient of air and water vapor ($m^{\wedge 2}.s^{\wedge(-1)}$)

J_V and J_A: molar flow density in the cross section of a pore, for water vapor and air (kmol m^{-2} s^{-1})

c_V and c_A: molar concentrations of water vapor and air (kmol m^{-3})

RT/D_{VP}: "friction coefficient" for molecules of water vapor in contact with the sides of the pores (kg $kmol^{-1}$ s^{-1})

In the Knudsen regime, where molecules do not interact with each other but only with the sides of the pores, the first term of the second member, which takes account of mutual friction between the air and the water, should be removed.

Furthermore, using the ideal gas hypothesis:

$$-\nabla\mu_V = -RT\nabla Lny_V$$

$$-\bar{V}_V\nabla P_T = \frac{-RT}{P_T}\nabla P_T = -RT\nabla LnP_T$$

(P_T is the total pressure of the gaseous phase).

The first member becomes:

$$-\nabla\mu_V - \bar{V}_V\nabla P_T = -RT\nabla LnP_T y_V = -RT\nabla LnP_V$$

Using membrane theory, we may write:

$$\frac{RT}{D_{VP}} = \frac{\eta_V \bar{V}_V}{K_V}$$

η_V: viscosity of water vapor (Pa s)

K_V: permeability of the pore for water vapor (m^2)

However, we have seen that:

$$\eta_V = \frac{1}{3}\rho_V v_V^* \lambda \quad \text{and} \quad K_V = K_0 F = \frac{R_0^2}{8} \times \frac{8\lambda}{R_0}\sqrt{\frac{2}{\pi}}$$

Furthermore:

$$\bar{V}_V = \frac{RT}{P_T} \quad \text{and} \quad \rho_V = \frac{P_T}{RT} M_V y_V$$

Hence:

$$\frac{RT}{D_{VP}} = \frac{1}{3}\sqrt{\frac{\pi}{2}} \frac{M_V y_V v_V^*}{R_0}$$

The molar concentration of the vapor is obviously:

$$c_V = \frac{P_T y_V}{RT}$$

This gives us the vapor migration equation for the Knudsen regime:

$$J_V = -3\sqrt{\frac{2}{\pi}} \frac{P_T R_0}{M_V v_V^*} \nabla \text{Ln} P_V \qquad \left(\text{kmol.m}^{-2}.\text{s}^{-1}\right)$$

The molar flow density now needs to be considered in relation to the free surface of the product, rather than the straight section of a pore. For this, we use:

– the porosity ε, which is the volume fraction of the product occupied by pores; this is also the fraction of a section of product occupied by pores;

– the tortuosity of the pores, which is the relationship between the actual length of the pores and the thickness of the product.

The molar flow density referred to the surface of the product is:

$$N_V = J_V \frac{\varepsilon}{t^2}$$

This quantity is proportional to ε as the gas flow is itself proportional to the total of the pore sections.

The tortuosity value is squared, as it has two effects:

– dividing the pressure gradient by extending the path;

– multiplying the flow density in each pore, as, for a constant pore radius and constant porosity, the number of pores is inversely proportional to t.

Finally:

$$N_V = -\frac{3\varepsilon}{t^2}\sqrt{\frac{2}{\pi}\frac{P_T R_0}{M_V v_V^*}}\nabla \mathrm{Ln} P_V$$

Note that ∇ is the "nabla" sign, a partial derivative in relation to the space variables.

17.3.5. *Mean pore radius*

The expression of the mean pore radius R_0 can be obtained using classic liquid filtration theory (see [DUR 99]).

$$R_0 = \frac{2\varepsilon}{(1-\varepsilon)\sigma}$$

σ is the contact surface between the liquid and the solid per m^3 of actual solid.

For a freeze-dried, i.e. dry, solid, it is possible to measure ε_0 (by weighing) and σ_0 (by percolation, using a liquid other than water).

For a product during the freeze-drying process, the porosity ε is less than ε_0, and we can write:

$$\sigma = \frac{6}{d} = \frac{6}{d_0}\left(\frac{1-\varepsilon_0}{1-\varepsilon}\right)^{1/3} = \sigma_0\left(\frac{1-\varepsilon_0}{1-\varepsilon}\right)^{1/3}$$

The speed of the initial freezing process has an influence on the pore radius.

If the initial freezing process is rapid, small ice crystals are formed; after sublimation, they leave pores with a small radius, the permeability of the material is low, and the vaporization of solvent water is slow. In other words, the dehydration speed at the end of the process is slow. The situation is reversed for a slow initial freezing process.

17.3.6. *Relationship between instantaneous porosity and water content*

At the end of the freeze-drying process, we obtain a porous body of porosity ε_0. The mass of solid dry matter per m^3 of product is thus:

$$\rho_S \times (1-\varepsilon_0)$$

ρ_S is the actual density of the solid.

After freezing but before freeze-drying, water is present in two different forms, and the quantity per kg of dry solid may be expressed using:

– Y kg of ice;

– X kg of solvent water.

The total mass of water per m^3 of product to freeze-dry is thus:

$$\rho_S (1-\varepsilon_0)(X+Y) \qquad [17.1]$$

The mass (X+Y) occupies a volume:

$$\frac{X}{\rho_L} + \frac{Y}{\rho_{GL}}$$

ρ_L and ρ_{GL} are the densities of solvent water and ice.

The density of the ice and water together is:

$$(X+Y) \times \left(\frac{X}{\rho_L} + \frac{Y}{\rho_{GL}} \right)^{-1}$$

The volume occupied by gas and water per m^3 of product being treated is $(\varepsilon_0 - \varepsilon)$, where ε is the instantaneous porosity.

The total mass of water per m^3 of product is thus:

$$(\varepsilon_0 - \varepsilon)(X + Y) \times \left[\frac{X}{\rho_L} + \frac{Y}{\rho_{GL}} \right]^{-1} \qquad [17.2]$$

Setting equations [17.1] and [17.2] as equal and simplifying by X+Y, we obtain:

$$(\varepsilon_0 - \varepsilon) = \left(\frac{X}{\rho_L} + \frac{Y}{\rho_{GL}} \right) \rho_S (1 - \varepsilon_0)$$

i.e.:

$$\varepsilon = \varepsilon_0 \left(1 + \frac{X\rho_S}{\rho_L} + \frac{Y\rho_S}{\rho_{GL}} \right) - \rho_S \left(\frac{X}{\rho_L} + \frac{Y}{\rho_{GL}} \right)$$

This gives an expression of instantaneous porosity as a function of the solvent water and ice contents, X and Y.

17.4. Simulation of freeze-drying

Once a product has been frozen, it is vacuum treated in order to sublime the ice.

In the frozen part of the product, heat is transmitted by simple conduction. In the de-iced part, however, we observe:

– heat transmission by conduction;

– vaporization of solvent water (where applicable).

Sublimation of the ice constitutes a further source of cold for the ice.

The frozen part of the product is bordered by an ice sublimation front, moving slowly from the free surface to the insulating wall.

The equations for the problem are not given here. Readers may wish to consult the procedures used for drying porous masses (see section 4.8.3 in [DUR 16]) and the results obtained for the freezing process (see Chapter 16).

APPENDICES

Appendix 1

Characteristics of Exchanger Tubes

Ø EXT.	BWG	Wall thickness mm	Internal Ø mm	Cross-section, S cm^2	External surface area $m^2.m^{-1}$	Internal surface area $m^2.m^{-1}$	Weight in $kg.m^{-1}$ Steel d: 7.85
1/2″	12	2.77	7.16	0.403	0.399	0.228	0.678
	14	2.101	8.48	0.565		0.267	0.551
	16	1.65	9.40	0.694		0.295	0.450
(12.7)	18	1.245	10.2	0.819		0.321	0.353
	20	0.89	10.92	0.936		0.343	0.259
3/4″	10	3.40	12.25	1.175	0.0598	0.385	1.312
	11	3.05	12.95	1.316		0.0407	1.203
	12	2.77	13.5	1.44		0.0425	1.113
	13	2.415	14.25	1.595		0.0477	0.985
	14	2.101	14.85	1.729		0.0466	0.877
(19.05)	15	1.83	15.4	1.865		0.0484	0.775
	16*	1.65	15.75	1.95		0.0495	0.708
	17	1.475	16.1	2.025		0.0506	0.639
	18	1.245	16.55	2.155		0.0521	0.548
1″	8	4.19	17.	2.29	0.0798	0.0535	2.195
	9	3.76	17.9	2.51		0.0562	2.001
	10	3.40	18.6	2.715		0.0584	1.844
	11	3.05	19.3	2.935		0.0607	1.680
	12	2.77	19.85	3.09		0.0625	1.548

	13	2.415	20.6	3.325		0.0647	1.361
(25.4)	14*	2.101	21.2	3.52		0.0665	1.206
	15	1.83	21.75	3.715		0.0683	1.060
	16	1.65	22.1	3.835		0.0694	0.966
	17	1.475	22.45	3.96		0.0706	0.870
	18	1.245	22.9	4.12		0.0720	0.744
1″1/4	8	4.19	23.35	4.29	0.0997	0.0735	2.852
	9	3.76	24.25	4.615		0.0762	2.588
	10	3.40	24.95	4.885		0.0784	2.376
	11	3.05	25.65	5.165		0.0807	2.158
	12	2.77	26.15	5.39		0.0824	1.998
	13	2.415	26.95	5.71		0.0847	1.736
(31.75)	14	2.101	27.45	5.96		0.0865	1.569
	15	1.83	28.2	6.19		0.0884	1.311
	16	1.65	28.45	6.36		0.0894	1.224
	17	1.475	28.7	6.52		0.906	1.136
	18	1.245	29.2	6.715		0.0920	0.958
1″1/2	8	4.19	29.7	6.94	0.1198	0.0935	3.509
	9	3.76	30.5	7.36		0.0960	3.213
	10	3.4	31.25	7.68		0.0983	2.927
	11	3.05	32	8.075		0.1006	2.635
	12	2.77	32.5	8.33		0.1023	2.436
(38.1)	13	2.415	33.3	8.72		0.1045	2.112
	14	2.101	33.75	9.04		0.1065	1.926
	15	1.83	34.55	9.3		0.1083	1.589
	16	1.65	34.8	9.49		0.1095	1.482
	17	1.475	35.05	9.68		0.1105	1.375
	18	1.245	35.55	9.94		0.1120	1.157

* Most commonly used tubes.

Appendix 2

Resistance, Conductance, Diffusance

1) For electricity, Ohm's law is written as:

$$\Delta V = Ri = i \times \left(\frac{\rho \Delta L}{S}\right) = \text{electrical flow through surface } S \times \text{ resistance}$$

The intensity i is measured in $C.s^{-1}$.

The resistance expression features:

– the resistivity, ρ (Ω m);

– the distance traveled, ΔL;

– the cross-section, S.

2) In heating technology, Fourier's law is written as:

$$q = \lambda \times \frac{\Delta T}{\Delta L} S \qquad \left(J.s^{-1} = \text{Watt}\right)$$

or:

$$\Delta T = q \times \left(\frac{\Delta L}{S\lambda}\right) = \text{heat flow through surface } S \times \text{ conductance}$$

The conductance expression features:

– the conductivity, λ (J m–1°C–1 s–1);

– the distance traveled, ΔL;

– the cross-section, S.

3) For the diffusion of materials, Fick's law is written as:

$$m = SD\frac{\Delta C}{\Delta L} \qquad \left(\text{kg s}^{-1} \text{ or kmol s}^{-1}\right)$$

or

$$\Delta C = m \times \left(\frac{\Delta L}{SD}\right) = \text{flow of matter through surface S} \times \text{ diffusance}$$

C: concentration: kg or kmol.m^{-3}

The diffusance expression features:

– diffusivity D ($m^2\ s^{-1}$);

– the distance traveled, ΔL;

– the cross-section, S.

In these three laws, we see that the temperature T is a thermal (i.e. heat) potential, and the concentration C is a material potential. V is evidently the electrical potential, i.e. the voltage (in Volts).

NOTE.–

The following grouping may be used in thermal calculations:

$$\alpha = \frac{\lambda}{\rho C_p} \text{ with dimensions } \frac{J}{s\, m°C} \times \frac{m^3}{kg} \times \frac{kg°C}{J} = \frac{m^2}{s}$$

Grouping α is referred to as thermal diffusivity, as it uses the same units as the material diffusivity D.

Bibliography

[ABD 85] ABDALLA H., SINGH R.P., "Simulation of thawing of foods using finite element method", *Journal of Food Process Engineering*, vol. 7, pp. 273–286, 1985.

[BEL 60] BELLS K.S., "Exchanger design based on the Delaware research program", *Petroleum Engineering*, vol. 32, pp. C26–C40, 1960.

[BER 67] BERGLES A.E., LOPINA R.F., FIORI M.P., "Critical heat flux and flow observations for low pressure water flowing in tubes", *Transactions of the American Society of Mechanical Engineers*, vol. 89, no. 1, p. 69, 1967.

[BIL 81] BILLET R., *Verdampfung und ihre technischen Anwendungen*, Verlag Chemie, 1981.

[BOW 40] BOWMAN R.A., MUELLER A.C., NAGLE W.M., "Mean temperature difference in design", *Transactions of the American Society of Mechanical Engineers*, vol. 62, pp. 283–294, 1940.

[BRI 63] BRIGGS D.E., YOUNG E.H., "Convection heat transfer and pressure drop of air flowing across triangular pitch banks of finned tubes", *Chemical Engineering Progress Symposium Series*, vol. 69, no. 41, pp. 1–10, 1963.

[BRO 46] BROWN G.P., DINARDO A., CHENG G.K. *et al.*, "The flow of gases in pipes at low pressures", *Journal of Applied Physics*, vol. 17, pp. 802–813, 1946.

[BRO 47] BROWN R.W., SCOTT R., TOYNE C., "An investigation of heat transfer in agitated jacketed cast iron vessels", *Transactions of the Institution of Chemical Engineers*, vol. 25, p. 181, 1947.

[BRO 59] BROOKS G., SU G.J., *Chemical Engineering Progress*, vol. 55, no. 10, p. 54, 1959.

[CAS 84a] CASTAIGNE F., *Ind. Alim. Agric.*, vol. 101, no. 6, pp. 499–505, 1984.

[CAS 84b] CASTAIGNE F., *Ind. Alim. Agric.*, vol. 101 no. 9, pp. 767–775, 1984.

[CAS 85] CASTAIGNE F., LACROIX C., "Méthode simple permettant d'estimer les temps de congélation de produits ayant des formes simples", *Revue Générale du Froid*, pp. 251–263, 1985.

[CHE 66a] CHEN J.C., "Correlation for boiling heat transfer to saturated fluids in convective flow", *Research Gate*, vol. 5, no. 3, pp. 322–329, 1966.

[CHE 66b] CHEN I.E.C., "A correlation for boiling heat transfer of saturated fluids in convective flow", *Industrial & Engineering Chemistry Process Design and Development*, vol. 5, no. 3, pp. 322–329, 1966.

[CHE 85] CHEN C.S., "Thermodynamic analysis of the freezing and thawing of foods: enthalpy and apparent specific heat", *Journal of Food Science*, vol. 50, pp. 1158–1166, 1985.

[CHI 33] CHILTON T.M., COLBURN A.P., GÉNÉRAUX R.P. *et al.*, *Transactions of the American Society of Mechanical Engineers*, vol. 5, 1933.

[CHO 71] CHOUMER S., Corrosion par les résidus de combustion du mazout lourd dans les chaudières industrielles, Thesis, CNAM, Paris, May 1971.

[CLE 84] CLELAND A.C., EARLE R.L., "Assessment of freezing time prediction methods", *Journal of Food Science*, vol. 49, pp. 1034–1042, 1984.

[CLE 86] CLELAND D.J., CLELAND A.C., EARLE R.L. *et al.*, "Prediction of thawing times for foods of simple shape", *International Journal of Refrigeration*, vol. 9, pp. 220–228, 1986.

[CLE 87a] CLELAND D.J., CLELAND A.C., EARLE R.L., "Prediction of freezing and thawing times for multi-dimensional shapes by simple formulae. Part I: regular shapes", *International Journal of Refrigeration*, vol. 10, pp. 156–164, 1987.

[CLE 87b] CLELAND D.J., CLELAND A.C., EARLE R.L., "Prediction of freezing and thawing times for multi-dimensional shapes by simple formulae. Part II: irregular shapes", *International Journal of Refrigeration*, vol. 10, pp. 234–240, 1987.

[CLE 87c] CLELAND D.J., CLELAND A.C., EARLE R.L. *et al.*, "Prediction of freezing and thawing times for multi-dimensional shapes by numerical methods", *International Journal of Refrigeration*, vol. 10, pp. 32–39, 1987.

[COX 72] COX C.C., DYER D.F., "Freeze-drying of spheres and cylinders", *Journal of Heat Transfer*, vol. C94, pp. 57–63, 1972.

[DE 80] DE ELVIRA C., DIAZ M., LOMBRANA I., "Modelo de poros al azar para la simulación de la liofilización de disoluciones congeladas con control termico", *Revista de Agroquimica y Tecnologia de Alimentos,* vol. 26, no. 1, pp. 148–156, 1980.

[DE 84] De Vriendt A.B., *La transmission de la chaleur*, vol. 2, Editions Gaétan Morin, 1984.

[DUC 59] Duckler A.E., "Dynamics of vertical falling film systems", *Chemical Engineering Progress*, vol. 55, pp. 62–67, 1959.

[DUR 99] Duroudier J.P., *Pratique de la filtration*, Editions Hermes, 1999.

[DUR 16] Duroudier J.P., *Adsorption - Dryers for Divided Solids*, ISTE Press Ltd, London and Elsevier Ltd, Oxford, 2016.

[DYE 66] Dyer D.F., Sunderland J.E., "Bulk and diffusional transport in the region between molecular and viscous flow", *International Journal of Heat and Mass Transfer*, vol. 9, pp. 519–526, 1966.

[ENG 59] Engelberg-Forster K., Greif R., "Heat transfer to a boiling liquid – mechanism and correlations", *Journal of Heat Transfer*, vol. 81, pp. 43–53, February 1959.

[FAI 60] Fair J.R., "What you need to design Thermosiphon reboilers", *Petroleum Refiner*, vol. 39, no. 2, pp. 105–123, 1960.

[FAI 63a] Fair J.R., "Vaporizer and reboiler design: part 1", *Chemical Engineering*, vol. 70, pp. 119, July 1963.

[FAI 63b] Fair J.R., "Vaporizer and reboiler design: part 2", *Chemical Engineering*, vol. 70, p. 101, August 1963.

[FEY 79] Feynman R.P., Leighton R.B., Sands M.L., *Cours de physique*, Intereditions, 1979.

[FIN 03] Fine F., Gervais P., "Décontamination des poudres alimentaires. Revue bibliographique et nouvelles perspectives", *Science des Aliments*, vol. 23, pp. 367–394, 2003.

[FLI 72] Flink J., Karel M., "Mechanisms of retention of organic volatiles in freeze-dried systems", *Journal of Food Technology*, vol. 7, pp. 199–211, 1972.

[FOR 55] Forster H.K., Zuber N., "Dynamics of vapor bubbles and boiling heat transfer", *AIChE Journal*, vol. 1, no. 4, pp. 531–535, December 1955.

[FOR 59] Forster K.E., Greilf R., "Heat transfer to a boiling liquid. Mechanism and correlations", *Journal of Heat Transfer*, vol. 81, pp. 43–53, 1959.

[GOU 65] Goudet G., *Les fonctions de Bessel et leurs applications en physique*, Masson, 1965.

[HAR 85] Harrison M.A., Bishop P.J., "Parametric study of economical energy usage in freezing tunnels", *Revue Internationale du Froid*, vol. 8, pp. 29–36, January 1985.

[HAS 73] HASTED J., *Aqueous Dielectrics*, Chapman & Hall, London, 1973.

[HAY 83a] HAYAKAWA K., NONINO C., SUCCAR J., "Two dimensional heat conduction in food undergoing freezing: predicting freezing time of rectangular or finitely cylindrical food", *Journal of Food Science*, vol. 48, pp. 1841–1848, 1983.

[HAY 83b] HAYAKAWA K., NONINO C., SUCCAR J. *et al.*, "Two dimensional heat conduction in food undergoing freezing. Development of computerized model", *Journal of Food Science*, vol. 48, p. 1849, 1983.

[HEI 71] HEILIGENSTAEDT W., *Thermique appliqué aux fours industriels*, Dunod, 1971.

[HEW 67] HEWIT B., KEEYS K., "Pulling studies of burnout in boiling heat transfer", *Transactions of the Institution of Chemical Engineers*, vol. 45, 1967.

[HOG 74] HOGG R., SHOJI K., AUSTIN L.G., "Axial transport of dry powders in horizontal rotating cylinders", *Powder Technology*, vol. 9, pp. 99–106, 1974.

[HOR 70] HORN R.K., Messungen zum Wärme und Stoffübegang am Rieselfilm, Karlsruhe University, 1970.

[HOT 67] HOTTEL H.C., SAROFIM A.F., *Radiative Transfer*, McGraw Hill, 1967.

[JEA 83] JEANNE M.P., "Technologie de la surgélation. Le materiel", *L'Alimentation et la Vie*, vols. 70–4, pp. 3–8, 1983.

[KER 50] KERN D.Q., *Process Heat Transfer*, McGraw Hill, 1950.

[KLI 54] KLINKENBERG A., "Heat transfer in cross-flow. Heat exchangers and packed beds", *Industrial and Engineering Chemistry*, vol. 46, no. 11, pp. 2285–2289, 1954.

[KUT 51] KUTATELADSE, Invest. Akad. Nauk. SSSR. Otdel. Teckh. Nauk. No. 4, p. 529, 1951.

[LED 38] LEDINEGG M., "Unstabilität der Strömung bei natürlichem und Zwangumlauf", *Warme*, vol. 61, p. 891, 1938.

[LEE 56] LEE D.C., DORSEY J.W., MOORE G.Z. *et al.*, "Design data for thermosiphon reboilers", *Chemical Engineering Progress*, vol. 52, p. 160, 1956.

[LEU 02] LEUSCHNER R.G.K., LILLFORD P.J., "Thermal properties of bacterial spores and biopolymers", *International Journal of Food Microbiology*, vol. 80, pp. 131–143, 2002.

[LEV 60] LEVINE D.G., FRIEDLANDER S.K., "The condensation of a vapour by mixing with a cool gas", *Chemical Engineering Science*, vol. 13, no. 2, pp. 49–56, 1960.

[LIA 79] LIAPIS A., LITCHFIELD R.J., "Optimal control of a freeze-dryer-I", *Chemical Engineering Science*, vol. 34, pp. 975–981, 1979.

[LIT 79] LITCHFIELD R.J., LIAPIS A.I., "An adsorption-sublimation model for a freeze-dryer", *Chemical Engineering Science*, vol. 34, pp. 1085–1090, 1979.

[LIT 82] LITCHFIELD R.J., LIAPIS A.J., "Optimal control of a freeze-dryer II", *Chemical Engineering Science*, vol. 17, pp. 45–55, 1982.

[LOC 49] LOCKHART R.W., MARTINELLI R.C., "Proposed correlation of data for isothermal two-phase, two component flow in pipes", *Chemical Engineering Progress*, vol. 45, no. 1, p. 39, January 1949.

[LON 69] LONCIN M., *Die grundlagen der Verfahrenstechnik in der Lebensmittelindustrie*, Sauerländer Aurau, Frankfurt, 1969.

[LON 85] LONCIN M., *Génie industriel alimentaire*, Masson, 1985.

[MAR 85] MARIN M., RIOS G.M., GIBERT H., "Use of time-temperature data during fluidized bed freezing to determine frozen food properties", *Journal of Food Process Engineering*, vol. 7, pp. 253–264, 1985.

[MAZ 99] MAZAS M., MARTINEZ S., LOPEZ M. *et al.*, "Thermal inactivation of bacillus cereus spores affected by the solutes used to control water activity of the heating medium", *International Journal of Food Microbiology*, vol. 53, pp. 61–67, 1999.

[MEH 86] MEHRA D.K., "Selecting evaporators", *Chemical Engineering* pp. 56–72, 3 February 1986.

[MIK 70] MIKIC B.B., ROHSENOW W.M., GRIFFITH P., "On bubbles growth rates", *International Journal of Heat and Mass Transfer*, vol. 13, pp. 657–666, 1970.

[MIL 84] MILLMAN M.J., LIAPIS A.I., MARCHELLO J.M., "Guidelines for the desirable operation of batch freeze-driers during the removal of free water", *Journal of Food Technology*, vol. 19, pp. 725–738, 1984.

[NAG 75] NAGATA S., *Mixing Principles and Applications*, John Wiley and Sons, 1975.

[NUS 11] NUSSELT W., "Der Wärmeübergang im Kreuzstrom", *Zeitschrift des Vereines Deutscher Ingenieure*, vol. 55, p. 2021, 1911.

[NUS 16] NUSSELT W., "Die Oberflächenkondensation des Wasserdampfes", *Zeitschrift des Vereins Deutcher Ingenieure*, vol. 60, pp. 541–546, 1916.

[NUS 27] NUSSELT W., "Die Theorie des Winderhitzers", *Zeitschrift des Vereins Deutcher Ingenieure*, vol. 71, no. 3, pp. 85–91, 1927.

[NUS 30] NUSSELT W., "Eine neue Formel für den Wärmedurchgang im Kreuzstrom", *Zeitschrift des Vereines Deuscher Ingenieure*, vol. 55, no. 12, pp. 417–422, 1930.

[NYK 79] NYKVIST W., DECAREAU R., "Microwave heat roasting", *Proceedings of the IMPI Symposium*, Waterloo, Canada, p. 69 – Paper 4.2, 1979, and *The Journal of Microwave Power*, no. 11, pp. 3–24, 1976.

[OLD 54] OLDSHUE J.Y., GRETTON A.T. "Helical coil heal transfer in mixing vessels", *Chemical Engineering Progress*, vol. 50, no. 12, p. 615, 1954.

[PAL 85] PALMIERI L., DE CINDIO B., DALL AGLIO G. *et al.*, "Analisi termica del processo di surgelamento di patate", *Inductria Conserve*, vol. 60, pp. 3–8, 1985.

[PER 07] PERRY R.H., *Chemical Engineers' Handbook*, McGraw Hill, 2007.

[PHA 85] PHAM Q.T., "Analytical method for predicting freezing times of rectangular blocks of foodstuffs", *Revue Internationale du Froid*, vol. 8, pp. 43–47, January 1985.

[PHA 86] PHAM Q.T., "Simplified equation for predicting the freezing time of foodstuffs", *Journal of Food Technology*, vol. 21, pp. 209–219, 1986.

[PHA 87] PHAM Q.T., "Calculation of bound water in frozen food", *Journal of Food Science*, vol. 52, no. 1, pp. 210–212, 1987.

[PHI 84] PHILIPPON J., "Méthodes de blanchiment-refroidissement des légumes destinés à la congélation", *Science des aliments*, vol. 4, pp. 523–550, 1984.

[PIE 79] PIERCE B.L., "Heat transfer colburn factor equation spans all fluid-flow regimes", *Chemical Engineering*, p. 113, 1979.

[PLA 41] PLANK, "Beitrage zur Berechning und Bewertung der Gefriergeschwindigkeit von Lebensmitteln", *Zeitschrift für die Gesamte Kalte Industrie*, vol. 3, pp. 1–24, 1941.

[PLA 74] PLASS L., MOLERUS O., "Simultane-Wärmeüebergangs-und Druckverlust-messungen an Feststoff/Flussigkeits Suspensionen in Vebergangsbereich vom homogenen zum heterogenen Suspensions Transport", *Chemie Ingénieur Technik*, vol. 46, no. 8, p. 355, 1974.

[PLE 52] PLESSET M.S., ZWICK S.A., "A nonsteady heat diffusion problem with spherical symmetry", *Journal of Applied Physics*, vol. 23, no. 1, pp. 95–99, January 1952.

[PLE 54] PLESSET M.S., ZWICK S.A., "The growth of vapor bubbles in superheated liquid", *Journal of Applied Physics*, vol. 25, no. 4, pp. 493–500, 1954.

[PON 67] PONTER A.B., DAVIES G.A., ROSS T.K. *et al.*, "The influence of mass transfer on liquid film breakdown", *International Journal of Heat and Mass Transfer*, vol. 10, pp. 349–359, 1967.

[REM 87] REMY J., "Modern freezing facilities", *Revue Internationale du Froid*, vol. 10, pp. 165–174, 1987.

[REY 85] REYNOSO R.O., CALCELO A., "Comparison between fixed and fluidized bed continuous pea freezers", *Revue Internationale du Froid*, vol. 8, pp. 109–115, March 1985.

[ROB 66] ROBINSON K.K., BRIGGS D.E., "Pressure drop of air flowing across triangular pitch banks of finned tubes", *Chemical Engineering Progress Symposium Series*, vol. 62, no. 64, pp. 177–184, 1966.

[ROB 75] ROBERTS A.G., SHAH K.O., *The Chemical Engineering*, pp. 745–750, December 1975.

[ROS 83] ROSSET M.R., ROUSSEL-CIQUARD N., "Décongélation des aliments", *L'Alimentation et la Vie*, vol. 70-4, pp. 39–56, 1983.

[SCH 45] SCHMIDT Th.E., "The heating capacity of finned tubes", *Bulletin de l'Institut international du froid*, Annexe G 5, 1945–1946.

[SCH 76] SCHWARTZBERG H.G., "Effective heat capacities for the freezing and thawing of food", *Journal of Food Science*, vol. 41, pp. 152–156, 1976.

[SCH 77] SCHMIDT D., VOGELPOHL A., "Untersuchung von Trennwirkung und Druckverlust eines Kreuzstrombodens", *Chemie Ingenieur Technik*, vol. 11, no. 8, pp. 464–466, 1977.

[SEA 51] SEAMAN W.C., "Passage of solids through rotary kilns", *Chemical Engineering Progress*, vol. 47, no. 10, pp. 508–514, 1951.

[SHE 75] SHENG T.R., PECK R.E., *AIChE Symposium Series*, vol. 73, no. 163, pp. 124–130, 1975.

[SHE 78] SHENOY A.V., MASHELKAR R.A., "Laminar natural convection heat transfer to a viscoelastic fluid", *Chemical Engineering Science*, vol. 33, pp. 769–776, 1978.

[SIS 76] SISSON B., "Loads for unit heaters", *Chemical Engineering*, p. 101, 5 July 1976.

[SPI 74] SPIEGEL M.R., *Formules et tables de mathématiques*, McGraw Hill, 1974.

[SPI 92] SPIEGEL M.R., *Formules et tables de mathématiques*, McGraw Hill, 1992.

[THU 89] THUÉRY J., *Les micro-ondes et leurs effets sur la matière*, Technique et Documentation, 1989.

[TO 78a] TO E., FLINK J.M., "Collapse, a structural transition in freeze dried carbohydrates. Part I", *Journal of Food Technology*, vol. 13, pp. 551–565, 1978.

[TO 78b] TO E., FLINK J.M., "Collapse, a structural transition in freeze dried carbohydrates. Part II", *Journal of Food Technology*, vol. 13, p. 567, 1978.

[TOO 71a] TOOR H.L., "Fog formation in boundary value problems", *AIChE Journal*, vol. 17, no. 1, pp. 5–14, 1971.

[TOO 71b] TOOR H.L., "Fog vaporization and condensation in boundary value problems", *Industrial and Engineering Chemistry Fundamental*, vol. 10, no. 1, pp. 121–131, 1971.

[TSV 81] TSVETKOV T.D., VULCHANOV N.L., "Toward the mathematical modeling of heat and mass transfer in vacuum freeze-drying", *Cryobiology*, vol. 18, pp. 155–165, 1981.

[VAN 78] VAN DER MEER Th.T., HOOGENDORN C.J., "Heat transfer coefficients for viscous fluids in a static mixer", *Chemical Engineering Science*, vol. 33, pp. 1277–1282, 1978.

[VUI 86] VUIKOVA C.F., MAIRAMIAN C.I., KEAUDOPOULO L.N., *Maslichnoe cirie i rastitelnie masla*, Oudk, 1986.

[WAL 67] WALLGREN B.O., "Heat transfer in liquid film flow", *Industrial and Engineering Chemistry Fundamentals*, vol. 6, no. 3, pp. 479–480, 1967.

[WEN 61] WEN C.-Y., MILLER E.N., "Heat transfer in solids-gas transport lines", *Industrial and Engineering Chemistry*, vol. 53, no. 1, pp. 51–53, January 1961.

[WIL 56] WILKE W., *Warmeübertragung in Rieselfilme*, VDI Forschungsheft 490 VDI Verlag, Düsseldorf, 1956.

[WIT 68] WITZE C.P., SHROCK V.E., CHAMBRE P.L., "Flow about a growing sphere in contact with a plane surface", *International Journal of Heat and Mass Transfer*, vol. 11, p. 1637, 1968.

Index

A, B, C

D, E, F

G, H, I

J, K, L

M, O, P

R, S, T

V, W